Mechanical Engineering Series

Series Editor
Francis A. Kulacki, Department of Mechanical Engineering, University of
Minnesota, Minneapolis, MN, USA

The Mechanical Engineering Series presents advanced level treatment of topics on the cutting edge of mechanical engineering. Designed for use by students, researchers and practicing engineers, the series presents modern developments in mechanical engineering and its innovative applications in applied mechanics, bioengineering, dynamic systems and control, energy, energy conversion and energy systems, fluid mechanics and fluid machinery, heat and mass transfer, manufacturing science and technology, mechanical design, mechanics of materials, micro- and nano-science technology, thermal physics, tribology, and vibration and acoustics. The series features graduate-level texts, professional books, and research monographs in key engineering science concentrations.

Amir A. Aliabadi

Turbulence

A Fundamental Approach for Scientists and Engineers

 Springer

Amir A. Aliabadi
University of Guelph
Guelph, ON, Canada

Mechanical Engineering Series
ISBN 978-3-030-95413-0   ISBN 978-3-030-95411-6   (eBook)
https://doi.org/10.1007/978-3-030-95411-6

This Springer imprint is published by the registered company Springer Nature Switzerland AG
The registered company address is: Gewerbestrasse 11, 6330 Cham, Switzerland

Homa, Fariborz, Reza, Laleh, Aban, and Herman

Preface

Turbulent flow is a ubiquitous phenomenon found in many natural and human-made systems from the core of the earth to the top of the atmosphere. It is also found in outer space, influencing how galaxies form and evolve. The most simple and curious mind can put an intuitive definition to turbulent flow. For instance, turbulent flow is commonly defined as flow that exhibits unpredictability (chaos), swirl (eddy or vorticity) on many scales, diffusion, and dissipation. Yet, providing an exact definition and developing the mathematical framework to describe turbulence can become complex and confusing. I have been studying the subject for more than two decades, yet many of my exposures to the topic have been disappointing. I have often found myself lost in the equations describing turbulence, neither completely understanding the mathematical rigor nor fully comprehending particular models that *attempt* to predict a system's evolution.

In my opinion, the first reason for my unsuccessful attempts was that the topic was always introduced to me without enough coverage of fundamental concepts. Too often the material was focused on practical aspects of a particular analysis. The second reason was that most tools I have been using to model turbulence were highly integrated and automated application software, thanks to the developers, enabling me to perform a complex simulation with selecting a few menu items and then pushing a few run buttons. While efficient, such application software have always masked the inner workings of the equations, approximations, and solution methods throughout a modelling exercise. The third reason was that the subject matter has always been difficult by its very nature. As Horace Lamb have said at a meeting of the British Association in London in 1932:

> I am an old man now, and when I die and go to Heaven there are two matters on which I hope for enlightenment. One is quantum electrodynamics and the other is the turbulent motion of fluids. And about the former I am really rather optimistic.

To overcome the shortcomings in my own understanding of the topic, I decided to write this book in support of a graduate course in turbulence suitable for majors in engineering and science. The book has been made as brief as possible, only covering fundamentals, but with a deep level of mathematical rigor. In addition, turbulence measurement techniques, modelling, simulation, and applications are discussed briefly to provide a comprehensive treatment of the subject. It is hoped that this fundamental approach will help students appreciate the topic at a more profound level and develop an understanding that can help them tackle more sophisticated turbulence problems encountered in science and engineering.

The mathematical notation discussed in this book closely follows the graduate textbook *Turbulent Flows* by professor *Stephen B. Pope* (2000). In addition, I have carried SI units throughout the text so students can see the connection among various flow variables of interest. In my opinion, meticulous handling of units accounts for half of the understanding of any physical system. In an ideal world, engineering and science students would not write down a symbol or number without expressing the associated units. Another gold standard in handling units is to always use positive and negative exponents for the components of compound units, instead of using fractions. For instance, the unit of velocity is better written down as $[\text{ms}^{-1}]$ than $[\text{m/s}]$.

The book is organized into five parts. Part I provides fundamentals of turbulence, main hypotheses, and analysis tools. Part II provides various measurement techniques used to study turbulent flows. Part III provides the modelling and simulation framework to study turbulent flows. Part IV provides brief applications of turbulence in engineering and sciences. Part V provides basic statistical, mathematical, and numerical tools in the study of the subject. In closing, I encourage and thank the readers to critique this text and provide me feedback and possibly corrections to the material studied here.

Guelph, Ontario, Canada   Amir A. Aliabadi
August 2021

Acknowledgments

I appreciate the assistance of the editorial and production teams at Springer Nature by Michael Luby, Brian Halm, Michelle Harding, Amrita Unnikrishnan, Meenahkumary Aravajy, Cynthia Pushparaj, and their other colleagues. I thank my family for their support while I spent countless hours to research, collect, and compose the material found in this text. I am thankful to Dr. Nima Atabaki, Department of Mechanical Engineering, University of British Columbia, whose teachings of experimental thermo-fluids helped me compose the material on turbulence measurement techniques. I am grateful to Rachel Maeve McLeod for proofreading the text and suggesting editorial corrections. Finally, I appreciate the patience of all the readers and their potential feedback toward improvement of the quality of the material.

Contents

Part III   Turbulence Modelling and Simulation

14   Introduction to Modelling and Simulation 191

15   Turbulent-Viscosity Models 195

16   Large-Eddy Simulation Models 211

17   Direct Numerical Simulation 231

18   Wall Models 235

Part I
Fundamentals

Chapter 1
Introduction

Abstract This chapter introduces the main concepts in the study of turbulence. Turbulent flows are defined as flows exhibiting chaotic property changes, swirl, diffusion, and dissipation. The concept of stability is introduced in dynamical systems, which are characterized by stable, conditionally stable, or unstable conditions. An eddy is defined as a spinning fluid structure that is created by turbulence. The concept of the energy cascade is introduced, which explains the transfer of turbulence kinetic energy from large to small scales of motion. Finally, the study of turbulence is grouped under three categories of discovery, modelling, and control.

1.1 Overview

In fluid dynamics, *turbulent flow* is a flow regime characterized by *chaotic property changes* with time (e.g. flow velocity, temperature, or concentration of a transported substance), *swirl* (eddy or vorticity) on many scales, *diffusion*, and *dissipation*. Turbulence is the dominating force in the physics and chemistry of the atmosphere and hydrosphere. It is ubiquitous in everyday life from airflow in cities and buildings to water flow in rivers and oceans. Even voice generation by humans is only possible by the turbulent flow of air through the vocal tract. Figure 1.1 shows the turbulent flow process in the formation of a cloud system, caused by chaotic property changes in velocity, temperature, pressure, and liquid water content. Yet turbulence is difficult to understand, measure, model, or simulate, and it has been considered as one of the unsolved problems of classical physics.

At a fundamental level, turbulence is caused by growth of instabilities in fluid motion, which typically occurs in many practical applications in the atmosphere or hydrosphere. Figure 1.2 shows the concept of stability for a ball moving on a curved surface subject to a gravitational field with gravitational acceleration pointing downward. Under the stable condition, the ball moves back to the initial position after any initial perturbation. Under the conditionally stable situation, the ball may or may not move back to the initial position, given the strength of the initial perturbation. Under the unstable condition, the ball will undoubtedly move away from the initial position, given any amount of initial perturbation. Fluid flows behave

Fig. 1.1 Clouds exhibit turbulence, which is evident from their shape as a result of chaotic property changes in velocity, temperature, pressure, and liquid water content; a tethered weather balloon shown in the centre of image attempting to measure properties of the atmospheric boundary layer

Fig. 1.2 Visual demonstration for the concepts of stable, conditionally stable, or unstable systems; movement of a ball on a curved surface subject to the gravitational field with gravitational acceleration pointing downward

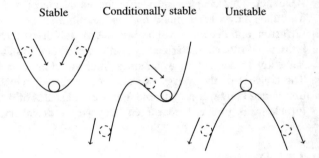

Stable   Conditionally stable   Unstable

similar to the conditionally stable situation, where small amounts of perturbation still permit orderly motion (laminar flow), while large amounts of perturbation initiate instability growth and chaos (turbulent flow).

The pioneering work in turbulence was conducted by Osborn Reynolds (1842–1912) [1–3]. He performed a series of experiments involving the injection of a dye at the centre of water flowing through a pipe. He observed that at some critical water velocity, the dispersion of the dye in the water transitions from laminar to turbulent regime. As was discovered by Reynolds, the transition from laminar to turbulent regimes was governed by the non-dimensional Reynolds number

$$Re = \frac{\mathcal{U}\mathcal{L}}{\nu} , \qquad (1.1)$$

where $\mathcal{U}$ [ms^{-1}] and $\mathcal{L}$ [m] are characteristic velocity and lengthscales of the flow, and ν [m^{2}s^{-1}] is the kinematic viscosity of the fluid. In physics, a *characteristic quantity* is a quantity that defines the scale of a physical system or phenomenon. For instance, the characteristic length defines the size of a system, and the characteristic velocity defines the velocity of flow through or around the system. The Reynolds number, defined by Eq. 1.1, quantifies the ratio of inertial to viscous forces. If $Re < 1000$, the flow is in the laminar range; if $1000 < Re < 4000$, the flow is in the transitional stage; and if $Re > 4000$, the flow is in the turbulent range. These limits for the Re number are provided for demonstration. In fact, these limits could be different based on the type of flow: internal flows (pipes, ducts, ...), external flows (atmospheric boundary layer, airfoils, ...), and more.

Turbulence can be perceived as the presence of eddies in the fluid [4]. An *eddy* is a local structure of a fluid that *spins*. Further, the *turbulent energy cascade* implies that turbulence begins with the formation of large eddies due to instability that subsequently break down to form smaller and smaller eddies due to subsequent instabilities until they are so small that they are damped by *viscous dissipation*. This phenomenon was observed in the early twentieth century and expressed poetically by Lewis Fry Richardson (1881–1953):

Big whirls have little whirls that feed on their velocity, and little whirls have lesser whirls and so on to viscosity.

The study of turbulence may fall under one of the three major categories: (i) *Discovery*: experimental or simulation studies providing qualitative or quantitative information about particular flows; (ii) *Modelling*: theoretical, modelling, or simulation studies aimed at developing rigorous mathematical models that can predict properties of turbulent flows; and (iii) *Control*: studies aimed at manipulating or controlling turbulent flow in a beneficial way.

Problems

1.1 For flow of air in a ventilation duct the characteristic velocity scale is $\mathcal{U} = 1 \text{ ms}^{-1}$ and the characteristic lengthscale, i.e. the hydraulic diameter of the duct, is $\mathcal{L} = 0.3 \text{ m}$. Assuming that the kinematic viscosity of air is $\nu = 1.5 \times 10^{-5} \text{m}^2\text{s}^{-1}$, calculate the Reynolds number of the flow and state if the flow is in the laminar or turbulent regime.

1.2 Provide an argument in favour of more aggressive instability growth in fluid motion when flow length and velocity scales are higher, and when dynamic viscosity of the fluid is lower. Does this demonstrate why level of turbulence in a flow is correlated with the Reynolds number?

References

1. Jackson J D (1995) Osborne Reynolds: scientist, engineer and pioneer. Proc R Soc Lond A 451:49–86
2. Jackson J D, Launder B (2007) Osborne Reynolds and the publication of his papers on turbulent flow. Annu Rev Fluid Mech 39:19–35
3. Launder B E, Jackson J D (2011) Osborne Reynolds: a turbulent life. In: Davidson P A, Kaneda Y, Moffatt K et al (ed) A Voyage Through Turbulence. Cambridge University Press, Cambridge
4. Richardson L F (1920) The supply of energy from and to atmospheric eddies. Proc R Soc Lond A 97:354–373

Chapter 2
Equations of Fluid Motion

Abstract This chapter introduces the transport equations of motion without the consideration of turbulence. Continuity, momentum, passive scalar, and vorticity transport equations are discussed. In the development of the transport equations, the concept of fluid element deformation is defined, where each parcel of fluid in the flow can be understood as being deformed via a combination of translation, linear strain, shear strain, and rotation mechanisms. Finally, the idea of similitude and non-dimensional transport equations is presented, where the transport equations are non-dimensionalized for characterizing their general behaviour. The concept of similarity for two systems involves geometric, kinematic, and/or dynamic similarities between them.

2.1 The Continuity Equation

The methodology developed in this chapter is adapted from [1]. For a flow with variable density ρ [kgm^{-3}] and velocity vector $\mathbf{U}$ [ms^{-1}] the continuity equation is given as

$$\underbrace{\frac{\partial \rho}{\partial t}}_{\text{Storage}} + \underbrace{\nabla.(\rho \mathbf{U})}_{\text{Advection}} = 0 , \qquad (2.1)$$

where t [s] is the time. If we assume ρ [kgm^{-3}] to be independent of space and time, i.e. to be constant, the continuity Eq. 2.1 simplifies to the kinematic condition that the velocity field should be *solenoidal* or *divergence-free*:

$$\nabla.\mathbf{U} = 0 . \qquad (2.2)$$

2.2 The Momentum Equation

The momentum equation in fluid dynamics is based on Newton's second law, which relates the fluid particle acceleration, i.e. the material derivative of fluid velocity $D\mathbf{U}/Dt$ [ms^{-2}], to the *surface forces* and *body forces* experienced by the fluid particle. In general, the surface forces are described by the *stress tensor* $\tau_{ij}(\mathbf{x}, t)$ [kgm^{-1}s^{-2}], with the property that it is symmetric ($\tau_{ij} = \tau_{ji}$). Here $\mathbf{x}$ indicates the position vector. The stress tensor is shown in Fig. 2.1.

The body force of interest is usually gravity. If Ψ [m^{2}s^{-2}] is the *gravitational potential* (i.e. the potential energy per unit mass associated with gravity), the body force per unit mass is

$$\mathbf{g} = -\nabla\Psi \,. \tag{2.3}$$

For a constant gravitational field the potential is $\Psi = gz$ [m^{2}s^{-2}], where g [ms^{-2}] is the gravitational acceleration and z [m] is the vertical coordinate. These forces cause the fluid particle to accelerate according to the momentum equation given by

$$\underbrace{\rho\frac{DU_j}{Dt}}_{\text{Material Derivative}} = \underbrace{\frac{\partial\tau_{ij}}{\partial x_i}}_{\text{Surface Force}} - \underbrace{\rho\frac{\partial\Psi}{\partial x_j}}_{\text{Body Force}} \,. \tag{2.4}$$

We now develop the momentum equation for flows of *constant-property Newtonian fluids*, e.g. constant density, the fundamental class of flows we consider in this

Fig. 2.1 Surface forces expressed as stress tensor components for a Cartesian fluid element. The stress tensor contains six shear components and three normal components

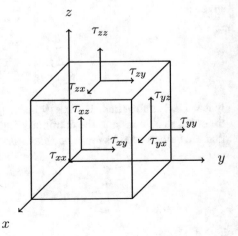

book. In this case, the stress tensor is

$$
\underbrace{\tau_{ij}}_{\text{Surface Force}} = - \underbrace{P\delta_{ij}}_{\text{Surface Normal Force}} + \underbrace{\mu\left(\frac{\partial U_i}{\partial x_j} + \frac{\partial U_j}{\partial x_i}\right)}_{\text{Surface Shear Force}} , \tag{2.5}
$$

where P [kgm^{-1}s^{-2}] is the pressure and μ [kgm^{-1}s^{-1}] is a constant coefficient of dynamic viscosity. By inserting this stress tensor, defined by Eq. 2.5, in the general momentum Eq. 2.4, and recalling that both ρ [kgm^{-3}] and μ [kgm^{-1}s^{-1}] are assumed constants and that $\nabla.\mathbf{U} = 0$, we obtain the Navier–Stokes equation for momentum

$$
\underbrace{\rho\frac{DU_j}{Dt}}_{\text{Material Derivative}} = \underbrace{\mu\frac{\partial^2 U_j}{\partial x_i \partial x_i}}_{\text{Surface Shear Force}} - \underbrace{\frac{\partial P}{\partial x_j}}_{\text{Surface Normal Force}} - \underbrace{\rho\frac{\partial \Psi}{\partial x_j}}_{\text{Body Force}} .
$$

$$\tag{2.6}$$

We can further compress this equation in shorter form by defining the *modified pressure p* [kgm^{-1}s^{-2}] by

$$
p = P + \rho\Psi , \tag{2.7}
$$

which simplifies the general momentum Eq. 2.6 to

$$
\underbrace{\frac{D\mathbf{U}}{Dt}}_{\text{Material Derivative}} = - \underbrace{\frac{1}{\rho}\nabla p}_{\text{Surface Normal and Body Forces}} + \underbrace{\nu\nabla^2\mathbf{U}}_{\text{Surface Shear Force}} ,
$$

$$\tag{2.8}$$

where $\nu = \mu/\rho$ [m^{2}s^{-1}] is the kinematic viscosity. Note that instead of Einstein's notation (see Chap. 23 for definition), this momentum Eq. 2.8 is now shown in vector form for velocity. In fact Eq. 2.8 represents the three momentum equations at once and is valid for each velocity component $U_1 = U$, $U_2 = V$, and $U_3 = W$ [ms^{-1}]. The Newtonian fluids with constant property are specified with Eq. 2.8 together with the solenoidal condition $\nabla.\mathbf{U} = 0$, which stipulates mass conservation.

In many practical applications, the viscous term containing μ [kgm^{-1}s^{-1}] is negligible compared to the pressure term containing p [kgm^{-1}s^{-2}]. Such flows are termed *inviscid flows*, in which the stress tensor is given as

$$
\tau_{ij} = -P\delta_{ij} . \tag{2.9}
$$

In this case the general momentum Eq. 2.8 is further simplified to the *Euler equation* given by

$$\underbrace{\frac{D\mathbf{U}}{Dt}}_{\text{Material Derivative}} = -\underbrace{\frac{1}{\rho}\nabla p}_{\text{Surface Normal and Body Forces}} , \qquad (2.10)$$

which does not contain the second partial derivative of velocity. Named after Leonhard Euler (1707–1783), the Euler Eq. 2.10 has applications in airfoil theory to calculate the lift force generated from an aerodynamic wing.

2.3 Conserved Passive Scalars

The Navier–Stokes Eq. 2.6 can be extended to express the transport of a *conserved passive scalar* denoted by $\phi(\mathbf{x}, t)$ [−]. In constant-property flows, the conservation equation for ϕ [−] is

$$\underbrace{\frac{D\phi}{Dt}}_{\text{Material Derivative}} = \underbrace{\Gamma\nabla^2\phi}_{\text{Diffusion}} , \qquad (2.11)$$

where Γ [m^{2}s^{-1}] can be a constant and known as uniform diffusivity. For a scalar quantity ϕ [−] to be passive, it must not appear in the generalized momentum Eq. 2.8, and so it cannot influence the fluid motion. It is termed *passive* since its value has no effect on material properties such as density, viscosity, and diffusivity. In other words, the flow will actually behave the same way whether or not there is a passive scalar present.

Usually the scalar quantity can represent various physical properties such as concentration or temperature. However, a passive scalar must only be present in small amounts, e.g. low concentrations or small excess temperatures, so that it does not affect the material properties in the flow. High concentrations of another substance in the flow or an excessive temperature difference causes density and viscosity variations. It must be noted that the Navier–Stokes Eq. 2.11 for scalars in general has more terms involved that are not discussed in the interest of the simplicity of the transport equations for our purposes in this book.

When the passive scalar is the temperature, the diffusivity is called the thermal diffusivity. The thermal diffusivity is non-dimensionalized using the kinematic viscosity in the so-called Prandtl number in honour of Ludwig Prandtl (1875–1953) by

$$Pr = \frac{\nu}{\Gamma} . \qquad (2.12)$$

When the passive scalar is the concentration of trace species, the diffusivity is called the molecular diffusivity. The molecular diffusivity is non-dimensionalized using the kinematic viscosity in the so-called Schmidt number in honour of Ernst Heinrich Wilhelm Schmidt (1892–1975) by

$$Sc = \frac{\nu}{\Gamma} . \tag{2.13}$$

For air at standard conditions, the Prandtl number, specified by Eq. 2.12, is in the range from 0.7 to 0.8. However, the Schmidt number, specified by Eq. 2.13, depends on the type of dilute mixture containing the vapor of a substance. For instance the Schmidt number can be as low as 0.22 for helium and as high as 2.66 for octane.

2.4 The Vorticity Equation

Turbulent flows contain eddies that by nature spin. As a result, turbulent flows are rotational, i.e. they have non-zero *vorticity*. The vorticity $\omega(\mathbf{x}, t)$ [s^{-1}] is defined as the curl of the velocity

$$\boldsymbol{\omega} = \nabla \times \mathbf{U} \tag{2.14}$$

and is equal to twice the rate of rotation of the fluid at any given space and time $(\mathbf{x}, t)$. The transport equation for vorticity can be obtained by taking the curl of the Navier–Stokes Eq. 2.8

$$\underbrace{\frac{D\omega}{Dt}}_{\text{Material Derivative}} = \underbrace{\nu\nabla^2\omega}_{\text{Diffusion}} + \underbrace{\omega.\nabla\mathbf{U}}_{\text{Vortex Stretching}} , \tag{2.15}$$

where the pressure term $(-\nabla \times \nabla p/\rho)$ in the Navier–Stokes Eq. 2.8 vanishes for constant-density flows. Vorticity has received much attention in turbulence research due to convenient properties. Unlike flow velocity, vorticity cannot be created nor destroyed within the interior of an ideal fluid and is rather transported within the flow by advective and diffusion processes. When we talk of a *turbulent eddy*, we really mean a blob of vorticity and its associated rotational and irrotational motion [2].

2.5 Fluid Element Deformation

Fluid elements experience deformation due to body and surface forces exerted on them. The deformation can be expressed as a combined effect of (a) translation, (b) linear strain, (c) shear strain, and (d) rotation as shown in Fig. 2.2 [3].

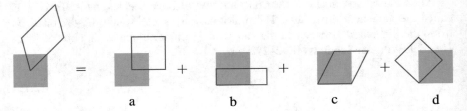

Fig. 2.2 Deformation of a fluid element in the Cartesian coordinates can be understood as the combined effects of (**a**) translation, (**b**) linear strain, (**c**) shear strain, and (**d**) rotation

These deformations can be expressed in terms of fluid velocity and velocity gradients. As noted earlier, the velocity gradients $\partial U_i / \partial x_j$ $[\text{s}^{-1}]$ are the components of a second-order stress tensor, defined by Eq. 2.5. For constant-density flow, the composition of $\partial U_i / \partial x_j$ $[\text{s}^{-1}]$ into *symmetric-deviatoric* and *antisymmetric* parts is [1]

$$\underbrace{\frac{\partial U_i}{\partial x_j}}_{\text{Deformation}} = \underbrace{S_{ij}}_{\text{Symmetric-deviatoric Deformation}} + \underbrace{\Omega_{ij}}_{\text{Antisymmetric Deformation}} .$$

(2.16)

S_{ij} $[\text{s}^{-1}]$ is the symmetric-deviatoric *rate-of-strain tensor*, and Ω_{ij} $[\text{s}^{-1}]$ is the antisymmetric *rate-of-rotation tensor*

$$S_{ij} \equiv \frac{1}{2} \left(\frac{\partial U_i}{\partial x_j} + \frac{\partial U_j}{\partial x_i} \right) , \tag{2.17}$$

$$\Omega_{ij} \equiv \frac{1}{2} \left(\frac{\partial U_i}{\partial x_j} - \frac{\partial U_j}{\partial x_i} \right) . \tag{2.18}$$

The rate-of-strain (Eq. 2.17) and rate-of-rotation (Eq. 2.18) tensors are very useful because transport equations can be expressed using them in shorter form. For instance, the Newtonian stress law can re-expressed Eq. 2.5 as

$$\tau_{ij} = -P\delta_{ij} + 2\mu S_{ij} , \tag{2.19}$$

indicating that the viscous stress depends linearly on the rate-of-strain independent of the rate-of-rotation. The vorticity and the rate-of-rotation are related by

$$\omega_i = -\epsilon_{ijk}\Omega_{jk} , \tag{2.20}$$

$$\Omega_{ij} = -\frac{1}{2}\epsilon_{ijk}\omega_k , \tag{2.21}$$

where ϵ_{ijk} [−] is the *alternating symbol*. Therefore, both vorticity (Eq. 2.20) and rate-of-rotation (Eq. 2.21) have the same information. However, while the rate-of-rotation is a tensor, vorticity is a vector.

2.6 Similitude and Non-Dimensional Transport Equations

A relevant question in fluid dynamics is whether or not two flow systems behave in the same fashion regardless of their size or properties. This is the concept of *similitude* in physics. Two systems may be similar in three ways. (1) *Geometric similarity* requires that the ratio of any two linear dimensions of two systems be the same. (2) *Kinematic similarity* is the similarity of time as well as geometry. It exists between two systems if the paths of moving particles are geometrically similar and if the ratios of the velocities of particles are similar. (3) *Dynamic similarity* exists between two systems when forces at corresponding points are similar. In fluid mechanics, to achieve dynamic similarity, all the non-dimensional numbers relevant to the flow must be preserved between the two systems, such as Reynolds, Grashof, and other numbers [4]. This is extremely limiting since in practice at most one or two non-dimensional numbers of the flow can be preserved.

Two geometrically similar systems may have different sizes, densities, and viscosities but the same velocity fields when they are appropriately scaled and referred to an appropriate coordinate system. Similar systems can be studied by the *transformation properties* of the Navier–Stokes equations. These properties are also called *invariance properties* or *symmetries*.

It can be shown that two flow systems are similar if they are geometrically similar and if they have the same non-dimensional transport equations. Consider a system with characteristic length $\mathcal{L}$ [m] and characteristic velocity $\mathcal{U}$ [ms^{-1}]. The characteristic length and velocity can be used to define the non-dimensional *independent variables*,

$$\hat{\mathbf{x}} = \frac{\mathbf{x}}{\mathcal{L}} , \tag{2.22}$$

$$\hat{t} = \frac{t\mathcal{U}}{\mathcal{L}} , \tag{2.23}$$

and also the *dependent variables* given the non-dimensional independent variables and system properties such as density,

$$\hat{\mathbf{U}}(\hat{\mathbf{x}}, \hat{t}) = \frac{\mathbf{U}(\mathbf{x}, t)}{\mathcal{U}} , \tag{2.24}$$

$$\hat{\mathrm{p}}(\hat{\mathbf{x}}, \hat{t}) = \frac{\mathrm{p}(\mathbf{x}, t)}{\rho\mathcal{U}^2} . \tag{2.25}$$

In applying these simple scaling transformations to the continuity and momentum Equations, i.e. Eqs. 2.2 and 2.8, respectively, we will obtain the non-dimensional continuity and momentum equations

$$\frac{\partial \hat{U}_i}{\partial \hat{x}_i} = 0 , \tag{2.26}$$

$$\frac{D\hat{U}_j}{D\hat{t}} = \frac{1}{Re} \frac{\partial^2 \hat{U}_j}{\partial \hat{x}_i \partial \hat{x}_i} - \frac{\partial \hat{p}}{\partial \hat{x}_j} , \tag{2.27}$$

where $Re = \mathcal{U}\mathcal{L}/\nu$ [−] is the familiar Reynolds number, also defined by Eq. 1.1. The non-dimensional transport equations are now only a function of the Re [−] parameter. In other words, these equations are identical as long as Re [−] is the same between two geometrically similar systems regardless of size or property differences. This ensures that the two flow systems be similar too.

Problems

2.1 A two-dimensional fluid velocity vector field is given as $\mathbf{U} = (x + y^2)\mathbf{i} + (x^2 + y)\mathbf{j}$ [ms^{-1}]. Show that the vorticity of the flow as a function of x and y [m] is given by

$$\boldsymbol{\omega} = (2x - 2y)\mathbf{k} . \tag{2.28}$$

2.2 For a simple shear flow near a wall, the one-dimensional fluid velocity vector field can be given as $\mathbf{U} = y\mathbf{i}$ [ms^{-1}], where unit vector $\mathbf{i}$ is the direction of flow along x axis parallel to the wall and unit vector $\mathbf{j}$ is the direction normal to the wall along the y axis. Show that for this flow the rate-of-strain tensor S_{ij} [s^{-1}] and the rate-of-rotation tensor Ω_{ij} [s^{-1}] (for $i = 1$ and $j = 2$) are given by

$$S_{ij} = \frac{1}{2} , \tag{2.29}$$

$$\Omega_{ij} = \frac{1}{2} , \tag{2.30}$$

suggesting that even for the simplest of shear flows there could be fluid element strain and rotation.

2.3 In a two-dimensional flow, a fluid is spinning around the vertical z axis in the direction of the unit vector $\mathbf{k}$. The flow field is given as $\mathbf{U} = -\Omega y\mathbf{i} + \Omega x\mathbf{j}$ [ms^{-1}], where Ω [s^{-1}] is a constant, also known as angular velocity. Here the unit vector $\mathbf{i}$ is the direction of flow along x axis and unit vector $\mathbf{j}$ is the direction of the flow along the y axis. Show that this flow is a *rotational flow*, i.e. a flow in which the vorticity is non-zero. In addition, show that for this flow the vorticity is equal to the following constant value in the entire flow domain, i.e.

$$\boldsymbol{\omega} = \nabla \times \mathbf{U} = 2\Omega\mathbf{k} . \tag{2.31}$$

2.4 In a two-dimensional flow, a fluid is spinning around the vertical z axis in the direction of the unit vector $\mathbf{k}$. The flow field is given as $\mathbf{U} = -\frac{y}{x^2+y^2}\mathbf{i} + \frac{x}{x^2+y^2}\mathbf{j}$ [ms^{-1}]. Here the unit vector $\mathbf{i}$ is the direction of flow along the x axis, and unit vector $\mathbf{j}$ is the direction of the flow along the y axis. Show that this flow is an *irrotational flow*, i.e. a flow in which the vorticity is precisely zero in the domain, i.e.

$$\boldsymbol{\omega} = \nabla \times \mathbf{U} = 0 \ . \tag{2.32}$$

2.5 In the previous problem show that the rate-of-rotation tensor is also identically equal to zero, hinting that for two-dimensional flows the rate-of-rotation tensor and vorticity give the same information, i.e.

$$\Omega_{ij} = 0 \ . \tag{2.33}$$

2.6 Verify the following relations:

$$\nabla.\boldsymbol{\omega} = 0 \ , \tag{2.34}$$

$$\nabla \times \nabla\phi = 0 \ , \tag{2.35}$$

$$\nabla \times (\nabla \times \mathbf{U}) = \nabla(\nabla.\mathbf{U}) - \nabla^2\mathbf{U} \ , \tag{2.36}$$

$$\mathbf{U} \times \boldsymbol{\omega} = \frac{1}{2}\nabla(\mathbf{U}.\mathbf{U}) - \mathbf{U}.\nabla\mathbf{U} \ . \tag{2.37}$$

2.7 A steady state velocity field in a fluid flow is given by $\mathbf{U} = -3x\mathbf{i} - 3y\mathbf{j} + 6z\mathbf{k}$ [ms^{-1}]. Show that the material derivative of this velocity field itself is given by

$$\frac{D\mathbf{U}}{Dt} = \frac{\partial\mathbf{U}}{\partial t} + \mathbf{U}.\nabla\mathbf{U} = 9x\mathbf{i} + 9y\mathbf{j} + 36z\mathbf{k} \ . \tag{2.38}$$

2.8 State if the following statement is true or false, why? *The material derivative of a scalar quantity is a scalar quantity, while the material derivative of a vector quantity is a vector quantity.*

2.9 The classical Kelvin–Stokes theorem relates the surface integral of the curl of a vector field $\mathbf{F}$ over a surface S in Euclidean three-dimensional space to the line integral of the vector field $\mathbf{F}$ over its boundary ∂S, which is a closed circuit enclosing S, i.e.

$$\iint_S (\nabla \times \mathbf{F}).d\mathbf{S} = \oint_{\partial S} \mathbf{F}.d\mathbf{r} \ . \tag{2.39}$$

In fluid dynamics, circulation is the line integral around a closed curve of the velocity field. Circulation is normally denoted by Γ [$m^2 s^{-1}$] and is shown as

$$\Gamma = \oint_{\partial S} \mathbf{U}.d\mathbf{l} . \tag{2.40}$$

Using the classical Kelvin–Stokes theorem show that the circulation and vorticity are related in such a way that circulation is the surface integral of the dot product of vorticity vector $\boldsymbol{\omega}$ [s^{-1}] with differential area vector $d\mathbf{S}$ [m^2] on a surface S enclosed by ∂S [m], i.e.

$$\Gamma = \oint_{\partial S} \mathbf{U}.d\mathbf{l} = \iint_S \boldsymbol{\omega}.d\mathbf{S} . \tag{2.41}$$

References

1. Pope S B (2000) Turbulent flows. Cambridge University Press, Cambridge
2. Davidson P A (2005) Turbulence: an introduction for scientists and engineers. Oxford University Press, Oxford
3. Acheson D J (1990) Elementary fluid dynamics. Oxford University Press, Oxford
4. Ettema R (2000) Hydraulic modelling: concepts and practice. American Society of Civil Engineering (ASCE), Reston

Chapter 3
Statistical Description of Turbulent Flows

Abstract This chapter characterizes turbulent flows from the perspective of statistics. The velocity field in the flow is defined as a random variable. The behaviour of this field can be characterized using a probability density function, which in turn can be described using its mean and moments. Further, in turbulent flows, more than one random variable may be considered simultaneously, whose relationship to other variables can be quantified using statistical tools developed for joint random variables, such as the joint probability density functions and correlation coefficients. One-point and two-point statistics (e.g. autocovariance, autocorrelation function, integral timescale, and integral lengthscale) are defined that help characterizing turbulent flows. The idea of Fourier transform pair is discussed, which enables transforming a turbulent signal from the time domain into the frequency domain and vice versa. Turbulent flows are categorized under statistically stationary, homogeneous, and/or axisymmetric conditions. Turbulent flows are further grouped under isotropic and/or anisotropic conditions. The idea of the wave number is introduced. Finally, types of averaging are discussed, which involve time averaging, ensemble averaging, and/or spatial averaging of a turbulent signal.

3.1 Preliminaries

The methodology developed in this chapter is adapted from [1]. In turbulent flow, the velocity field $\mathbf{U}(\mathbf{x}, t)$ [ms^{-1}] is random despite the fact that the Navier–Stokes equations, i.e. Eqs. 2.2 and 2.8, are deterministic, i.e. given the right boundary and initial conditions, the equations should predict an exact solution in a *prognostic* manner. How can this be? The answer lies in the combination of two observations: (1) in any turbulent flow there are unavoidable perturbations in initial conditions, boundary conditions, and material properties; (2) turbulent flow fields display an acute sensitivity to such perturbations [1]. These facts help explain why it is difficult to perform an exact prediction of turbulent flows even though the equations describing them are deterministic. An analogy is an experienced golf player who aims the ball on a target at a very far distance. However, the location of landing for the ball will exhibit errors due to minute initial misalignments in the club swing, imperfections in the club and the ball, and perturbations in air, e.g. winds, which all contribute to deviate the ball off target. Another point is that such minute misalignments, imperfections, and perturbations in air are never the same for two club swings, so every realization of the golf ball trajectory will be different under different conditions from one experiment to another.

For laminar flow conditions, the effects of perturbations in boundary conditions, initial conditions, and material properties are insignificant in creating instabilities; therefore, a measurement and prediction of flow using the Navier–Stokes equations will agree with a high degree of confidence. Navier–Stokes equations apply equally to turbulent flows, but the objective of the theory is different. In turbulent flows, since U [ms^{-1}] is a random variable, its value is unpredictable, so a theory cannot perform an exact prediction. However, a theory can either try to determine the probability of events that depend on the random variable U [ms^{-1}] or try to determine some statistical description of the random variable U [ms^{-1}], such as mean or standard deviation of the random variable.

3.2   Mean and Moments

The *mean* of the random variable U is defined by

$$\langle U \rangle \equiv \int_{-\infty}^{\infty} V f(V) dV , \qquad (3.1)$$

where the *probability density function (PDF)* $f(V)$ is the probability per unit distance in the sample space, hence the term probability density function. The PDF $f(V)$ has the dimensions of the inverse of U, while the $F(V)$, defined by Eq. 22.2, *cumulative distribution function (CDF)* and the product $f(V)dV$ are non-dimensional. Note that the mean as defined by Eq. 3.1 does not necessitate any averaging by time or space, but it simply defines the mean as variations of the random variable itself. $\langle U \rangle$ is the probability-weighted average of all possible values of U. More generally, if $Q(U)$ is any function of U, the mean of $Q(U)$ is given by

$$\langle Q(U) \rangle \equiv \int_{-\infty}^{\infty} Q(V) f(V) dV . \qquad (3.2)$$

If $Q(U)$ and $R(U)$ are two functions of the random variable U, the rules for taking means satisfy

$$\langle [aQ(U) + bR(U)] \rangle = a\langle Q(U) \rangle + b\langle R(U) \rangle . \qquad (3.3)$$

In other words, $\langle \rangle$ behaves as a linear operator. The *fluctuation* of U is defined by u as

$$u \equiv U - \langle U \rangle , \qquad (3.4)$$

and the *variance* is defined to be the mean square fluctuation by

$$\langle u^2 \rangle = \int_{-\infty}^{\infty} (V - \langle U \rangle)^2 f(V) dV \ . \tag{3.5}$$

The square-root of the variance is the *standard deviation* and denoted by $\langle u^2 \rangle^{1/2}$. In many textbooks the fluctuations are also shown as u'. The square-root of variance is also shown by σ_u. In general the nth *central moment* is defined as

$$\mu_n \equiv \langle u^n \rangle = \int_{-\infty}^{\infty} (V - \langle U \rangle)^n f(V) dV \ . \tag{3.6}$$

3.3 Standardization

It is possible to standardize random variable fluctuations. The standardized random variable fluctuation has zero mean and unit variance. The standardized random variable $\hat{U}$ corresponding to U is

$$\hat{U} \equiv (U - \langle U \rangle)/\sigma_u \ , \tag{3.7}$$

and its PDF, i.e. the standardized PDF of U, is

$$\hat{f}(\hat{V}) = \sigma_u f(\langle U \rangle + \sigma_u \hat{V}) \ . \tag{3.8}$$

The moments of $\hat{U}$, the standardized moments of U, are

$$\hat{\mu}_n = \frac{\langle u^n \rangle}{\sigma_u^n} = \frac{\mu_n}{\sigma_u^n} = \int_{-\infty}^{\infty} \hat{V}^n \hat{f}(\hat{V}) d\hat{V} \ . \tag{3.9}$$

Evidently, we have $\hat{\mu}_0 = 1$, $\hat{\mu}_1 = 0$, and $\hat{\mu}_2 = 1$. In statistics, the third standardized moment, $\hat{\mu}_3$, is called *skewness*. Skewness is a measure of the asymmetry of the probability distribution of a real-valued random variable about its mean. The skewness value can be positive or negative. The fourth standardized moment, $\hat{\mu}_4$, is called *kurtosis* or *flatness*. Kurtosis is a measure of the *tailedness* of the probability distribution of a real-valued random variable. In a similar way to the concept of skewness, kurtosis is a descriptor of the shape of a probability distribution [2, 3].

3.4 Joint Random Variables

The concept of a single random variable U can be extended to two or more random variables. For instance we can take the three components of a velocity vector (U_1, U_2, U_3) representing three random variables. A *joint probability density*

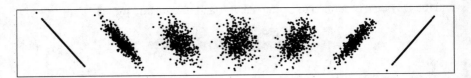

Fig. 3.1 Demonstration of correlation coefficient between two random variables with $\rho_{12} = -1, -0.9, -0.5, 0.0, +0.5, +0.9, +1.0$ from left to right

function (JPDF) can represent the density function for a pair of two random variables, such as $f_{12}(V_1, V_2)$. The *covariance* of U_1 and U_2 is the mixed second moment and can be expressed using the JPDF as

$$\langle u_1 u_2 \rangle = \int_{-\infty}^{\infty} \int_{-\infty}^{\infty} (V_1 - \langle U_1 \rangle)(V_2 - \langle U_2 \rangle) f_{12}(V_1, V_2) dV_1 dV_2 , \qquad (3.10)$$

and the *correlation coefficient* is

$$\rho_{12} \equiv \frac{\langle u_1 u_2 \rangle}{[\langle u_1^2 \rangle \langle u_2^2 \rangle]^{1/2}} . \qquad (3.11)$$

As shown in Fig. 3.1 a positive correlation coefficient occurs when a positive fluctuation of one random variable is most likely accompanied with a positive fluctuation of the other random variable as well (the same can be said for a negative fluctuation). A negative correlation coefficient occurs when a positive fluctuation of one random variable is most likely accompanied with a negative fluctuation of the other random variable. In Fig. 3.1 each axis represents one of the random variables, and the mean for both random variables occurs at the centre of the scatter plot.

3.5   Normal and Joint-Normal Distributions

The *central-limit theorem* and *Gaussian distribution* function play important roles in the theory of probability and that of turbulent flows. It is necessary to begin the theory by the concept of *ensemble average*. Assume U is a component of velocity at a particular space and time in a repeatable turbulent flow experiment and that $U^{(n)}$ denotes the nth repetition of the experiment, i.e. coming from another measurement of U at the same location and time. The set of random variables $\{U^{(1)}, U^{(2)}, U^{(3)}, \ldots, U^{(N)}\}$ are independent and have the same distribution, i.e. that of U, and they are said to be *independent and identically distributed*. The *ensemble average* is calculated by

$$\langle U \rangle_N \equiv \frac{1}{N} \sum_{n=1}^{N} U^{(n)} . \qquad (3.12)$$

Note that an ensemble average results from a finite number of measurements and is different from $\langle U \rangle$ given by Eq. 3.1. The ensemble average itself is a random variable. It is straightforward to show that its mean and variance are given as

$$\langle \langle U \rangle_N \rangle = \langle U \rangle \,, \tag{3.13}$$

$$\langle (\langle U \rangle_N - \langle U \rangle)^2 \rangle = \frac{\sigma_u^2}{N} \,. \tag{3.14}$$

So even though the means of the ensemble average and the random variable are the same, the standard deviation for an ensemble average reduces compared to the standard deviation for the random variable as the number of measurements increases. In the limit of $N \rightarrow \infty$, if the ensemble average is calculated over a very large sample, then its standard deviation approaches zero.

It is possible to standardize the ensemble average random variable fluctuations. Again, the standardized random variable has zero mean and unit variance. The standardized random variable $\hat{U}$ corresponding to $\langle U \rangle_N$ is defined by

$$\hat{U} = [\langle U \rangle_N - \langle U \rangle] N^{1/2} / \sigma_u \,. \tag{3.15}$$

The *central-limit theorem* states that, as $N \rightarrow \infty$, the PDF of $\hat{U}$, i.e. $\hat{f}(V)$, tends to the standardized *Gaussian distribution* or *normal distribution*

$$\hat{f}(V) = \frac{1}{\sqrt{2\pi}} \exp\left(-\frac{1}{2} V^2 \right) \,. \tag{3.16}$$

This is an important result, indicating that regardless of the nature of fluctuations, i.e. they could be expressed using any other PDF, the standardized fluctuations of the ensemble-averaged random variable will have a Gaussian or normal PDF.

Other ensemble-averaged joint random variables also exhibit joint Gaussian or normal distribution in homogeneous turbulence. For instance, the velocity components and a conserved passive scalar are found to be joint-normally distributed.

3.6   Random Processes

In many instances, particularly when observing natural phenomena such as atmospheric or oceanic flows, it is not possible to repeat an experiment in order to obtain ensemble averages by Eq. 3.12. In those cases, it is useful to develop theories for the same experiment but only as a function of time, for instance velocity as a function of time $U(t)$ [ms^{-1}]. Such a time dependent random variable is called a *random process*. Many random processes are *statistically stationary*. A process is statistically stationary if all multi-time statistics are invariant under a shift in time. For instance, in a statistically stationary atmosphere, if one performs a five-minute average of wind speed at one particular time, one will obtain the same average as another five-minute sample taken at another time. A turbulent flow, after an initial transient period, can reach a statistically stationary state, where even though the flow variables vary with time, the statistics are independent of time.

For a statistically stationary process, the simplest multi-time statistic that can be considered is the *autocovariance* given by

$$R(s) \equiv \langle u(t)u(t+s) \rangle , \qquad (3.17)$$

where a flow quantity is measured at the same location with a time shift s [s] over an indefinite time period. Note that autocovariance is not a function of time t [s] but instead a function of time shift s [s]. In the normalized form, this is called the *autocorrelation function* defined by

$$\rho(s) \equiv \frac{\langle u(t)u(t+s) \rangle}{\langle u(t)^2 \rangle} , \qquad (3.18)$$

where $u(t) \equiv U(t) - \langle U(t) \rangle$ [ms^{-1}] is the velocity fluctuation. The autocorrelation function is the correlation coefficient between a function and itself at time t and $t+s$ [s]. Consequently, $\rho(0) = 1$ [−] and $\rho(s > 0) \leq 1$.

Related to the concept of autocorrelation is the concept of *integral timescale*. Usually, the autocorrelation function diminishes quickly as a function of time shift. This means that usually the integral of the autocorrelation function (Eq. 3.18) over time shift converges to the integral timescale defined as

$$\overline{\tau} = \int_0^\infty \rho(s)ds . \qquad (3.19)$$

In essence, the integral timescale looks at the overall memory of the process and how strongly it is influenced or correlated by state of the flow in a previous time. A vivid example is a person standing in gusty winds. The timescale for person's sensation of each turbulent eddy arriving at the face is determined by the integral timescale. If the memory of the turbulent flow is short, the person will feel very quick bursts of eddies at the face where the integral timescale is short. If the memory of the turbulent flow is longer, the person will feel longer bursts of eddies at the face where the integral timescale is longer.

The autocovariance $R(s) \equiv \langle u(t)u(t+s) \rangle = \langle u(t)^2 \rangle \rho(s)$ [m^{2}s^{-2}] and twice the *frequency spectrum* $E(\omega)$ [m^{2}s^{-1}] form a *Fourier transform pair*:

$$
\begin{aligned}
E(\omega) &\equiv \frac{1}{\pi} \int_{-\infty}^{\infty} R(s)e^{-i\omega s} ds \\
&= \frac{2}{\pi} \int_0^{\infty} R(s) \cos(\omega s)ds ,
\end{aligned}
\qquad (3.20)
$$

and

$$R(s) \equiv \frac{1}{2} \int_{-\infty}^{\infty} E(\omega)e^{i\omega s}\,d\omega$$

$$= \int_{0}^{\infty} E(\omega)\cos(\omega s)\,d\omega\,, \tag{3.21}$$

where $\omega = 2\pi f$ [Rads^{-1}] is the frequency, calculated from frequency f [s^{-1}] in cycles per seconds. Both $R(s)$ [m^{2}s^{-2}] and $E(\omega)$ [m^{2}s^{-1}] contain the same information about the flow but in different forms. The Fourier transform pair (Eqs. 3.20 and 3.21) can be better understood using Euler's (1707–1783) formula

$$e^{ix} = \cos(x) + i\sin(x)\,, \tag{3.22}$$

which in our case can be re-written as

$$e^{i\omega s} = \cos(\omega s) + i\sin(\omega s)\,. \tag{3.23}$$

Note that since the integration is performed from $-\infty$ to ∞, the $i\sin(\omega s)$ term, which is an odd function, is integrated to zero and only the term $\cos(\omega s)$ remains, which is an even function. Due to this, both $R(s)$ [m^{2}s^{-2}] and $E(\omega)$ [m^{2}s^{-1}] (Eqs. 3.20 and 3.21) are even functions.

A fundamental property of the frequency spectrum is that for a range of frequencies $\omega_a < \omega < \omega_b$ [Rads^{-1}] the integral

$$\int_{\omega_a}^{\omega_b} E(\omega)\,d\omega \tag{3.24}$$

is the contribution to the variance $\langle u(t)^2 \rangle$ [m^{2}s^{-2}] of all modes in the frequency range $\omega_a < \omega < \omega_b$ [Rads^{-1}]. In other words, this integral gives the amount of variance $\langle u(t)^2 \rangle$ [m^{2}s^{-2}] due to turbulent fluctuations in that particular frequency range. Of course, the rest of the total amount of variance comes from turbulent fluctuations with frequencies outside this range. Given the frequency spectrum, it is easily possible to calculate the variance by

$$R(0) = \langle u(t)^2 \rangle = \int_{0}^{\infty} E(\omega)\,d\omega\,. \tag{3.25}$$

Another relationship can be established between the frequency spectrum and the autocorrelation function. It can be shown that

$$\overline{\tau} = \frac{\pi E(0)}{2\langle u(t)^2 \rangle}\,. \tag{3.26}$$

Not all random processes are differentiable, but random processes that arise from turbulent flows are differentiable so that by definition we have

$$\frac{dU(t)}{dt} = \lim_{\Delta t \to 0} \left(\frac{U(t + \Delta t) - U(t)}{\Delta t} \right) . \tag{3.27}$$

It is also possible to calculate the mean for the differentiation of a random process, realizing that the mean and taking the limit commute, so that

$$\left\langle \frac{dU(t)}{dt} \right\rangle = \left\langle \lim_{\Delta t \to 0} \left(\frac{U(t + \Delta t) - U(t)}{\Delta t} \right) \right\rangle$$

$$= \lim_{\Delta t \to 0} \left(\frac{\langle U(t + \Delta t) \rangle - \langle U(t) \rangle}{\Delta t} \right)$$

$$= \frac{d \langle U(t) \rangle}{dt} . \tag{3.28}$$

This result is very important and reads as *the mean of the derivative of a random process is equal to the derivative of the mean of the random process*. This result is extremely useful for deriving transport equations for turbulent flows.

3.7 Random Fields

So far we discussed a random process at a particular location in a turbulent flow. It is possible to extend the concept of random process to a region of space or field as well. The velocity $\mathbf{U}(\mathbf{x}, t)$ [ms^{-1}] is a time dependent random *vector field*. The fluctuating *velocity field* is defined by

$$\mathbf{u}(\mathbf{x}, t) \equiv \mathbf{U}(\mathbf{x}, t) - \langle \mathbf{U}(\mathbf{x}, t) \rangle . \tag{3.29}$$

The one-point and one-time covariance of the velocity is $\langle u_i(\mathbf{x}, t) u_j(\mathbf{x}, t) \rangle$ [m^{2}s^{-2}]. This covariance is computed for every spatial location in the flow field. This covariance is not to be confused with autocovariance (Eq. 3.17) and the autocorrelation function (Eq. 3.18), where a time shift was considered at the same spatial point. Rather this covariance is calculated at the same time for different components of the velocity in the same spatial location, for instance the horizontal and vertical components. These covariances for the velocity field are called *Reynolds stresses* and are written as $\langle u_i u_j \rangle$ [m^{2}s^{-2}], with dependencies on $\mathbf{x}$ [m] and t [s] being understood and the notation compressed without writing $\mathbf{x}$ [m] and t [s].

Turbulent velocity fields are also differentiable with respect to time and space, and the mean and the differentiation commute so that

$$\left\langle \frac{\partial U_i}{\partial t} \right\rangle = \frac{\partial \langle U_i \rangle}{\partial t} \, , \tag{3.30}$$

$$\left\langle \frac{\partial U_i}{\partial x_j} \right\rangle = \frac{\partial \langle U_i \rangle}{\partial x_j} \, . \tag{3.31}$$

3.8   Statistically Stationary, Homogeneous, and Axisymmetric Turbulent Flows

The random field $\mathbf{U}(\mathbf{x}, t)$ $[\text{ms}^{-1}]$ is *statistically stationary* if all statistics are invariant under a shift in time. Similarly, the field is *statistically homogeneous* if all statistics are invariant under a shift in position. Finally, the field is *statistically axisymmetric* if all statistics are independent of the circumferential coordinate. While the statistically stationary or axisymmetric conditions are more common to reach, the statistically homogeneous condition occurs less frequently in applied cases, unless considering a finite region of space. For instance, most turbulent flows are statistically stationary if boundary conditions are not time varying. For example, flow over a test article in a wind tunnel exhibits the statistically stationary condition. Most circular jets are statistically axisymmetric if there is no body force, such as gravity, exerted normal to the axis of the flow. For instance, a round nozzle sprinkler that ejects water upward is statistically axisymmetric. The statistically homogenous condition may be created in laboratory experiments or within a subset of flow region far away from boundaries.

3.9   Isotropic and Anisotropic Turbulence

A statistically homogeneous field $\mathbf{U}(\mathbf{x}, t)$ $[\text{ms}^{-1}]$ is, by definition, statistically invariant under translation. If the field is also invariant under rotations and reflections of the coordinate system, then it is *statistically isotropic*. For instance, in a statistically isotropic field $\langle u_1 u_2 \rangle = \langle u_1 u_3 \rangle = \langle u_2 u_3 \rangle$ $[\text{m}^2 \text{s}^{-2}]$ since the same Reynolds stresses must be obtained regardless of the rotational orientation of the coordinate system. Otherwise, the field is *statistically anisotropic*. Most of the turbulent theory is centred around this concept. In fact, many turbulence models assume a statistically isotropic condition to be valid. However, most turbulent flows deviate from the isotropic condition. For instance, the atmosphere is found to be anisotropic near the earth surface, i.e. in the atmospheric boundary layer [4–7]. Fortunately, many turbulence models are developed to treat such cases.

3.10   Two-Point Correlation

Up to here we discussed the one-point and two-time autocovariance $R(s)$ [m²s⁻²] $[\text{m}^2\text{s}^{-2}]$
(Eq. 3.17), where covariance in one location of the flow was calculated as a
function of a time shift s [s]. It is also possible to define a two-point and one-time
autocovariance defined by

$$R_{ij}(\mathbf{r}, \mathbf{x}, t) \equiv \langle u_i(\mathbf{x}, t)u_j(\mathbf{x} + \mathbf{r}, t)\rangle \,, \qquad (3.32)$$

which is also known as a *two-point correlation*. With this statistic it is possible to
define an *integral lengthscale* by

$$L_{11}(\mathbf{x}, t) \equiv \frac{1}{R_{11}(0, \mathbf{x}, t)} \int_0^\infty R_{11}(\mathbf{e}_1 r, \mathbf{x}, t)dr \,, \qquad (3.33)$$

where $\mathbf{e}_1$ is the unit vector in the x_1-coordinate direction. The integral lengthscale
measures the correlation distance of a process in terms of space or time. In essence,
it looks at the overall memory of the process and how it is influenced by previous
positions and parameters. An intuitive example would be the case in which you have
a very low Reynolds number flow such as a laminar flow, where the flow is fully
reversible and thus fully correlated with previous particle positions. This concept
may be extended to turbulence, where it may be thought of as the time during which
a particle is influenced by its previous position.

3.11   Wavenumber Spectra

We begin this section by defining the concept of *wavenumber vector*. The wavenumber is the spatial frequency of a wave, either in cycles per unit distance or radians per
unit distance. It can be envisaged as the number of waves that exist over a specified
distance (analogous to frequency being the number of cycles or radians per unit
time). The wave number can be defined by

$$|\kappa| = \frac{1}{\lambda} \,, \qquad (3.34)$$

$$|\kappa| = \frac{2\pi}{\lambda} \,, \qquad (3.35)$$

where λ [m] is the wavelength in the direction of the wavenumber vector κ [m⁻¹
or Radm⁻¹], with the first definition in units of cycles per unit distance and
the second definition in units of radians per unit distance. Figure 3.2 shows the

Fig. 3.2 Visual
representation of wave
propagation and the
wavelength

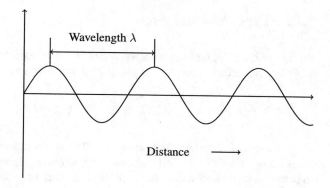

visual representation of the wave propagation and the wavelength. Note that the
wavenumber vector will be in the same direction as the distance vector shown.

Via Eqs. 3.20 and 3.21 we established a relationship between the frequency spectrum $E(\omega)$ [m^{2}s^{-1}] and the one-point and two-time autocovariance $R(s)$ [m^{2}s^{-2}].
This relationship was in the form of a *Fourier transform pair*. In this section we
seek a similar relationship but between the two-point and one-time autocovariance
$R_{ij}(\mathbf{r}, \mathbf{x}, t)$ [m^{2}s^{-2}] (Eq. 3.32) and velocity spectrum tensor $\Phi_{ij}(\kappa, t)$ [m^{5}s^{-2}],
which is a function of *wavenumber vector* κ,

$$\Phi_{ij}(\kappa, t) = \frac{1}{(2\pi)^3} \iiint_{-\infty}^{\infty} e^{-i\kappa \cdot \mathbf{r}} R_{ij}(\mathbf{r}, t) d\mathbf{r} , \qquad (3.36)$$

and

$$R_{ij}(\mathbf{r}, t) = \iiint_{-\infty}^{\infty} e^{i\kappa \cdot \mathbf{r}} \Phi_{ij}(\kappa, t) d\kappa , \qquad (3.37)$$

where $d\mathbf{r} = dr_1 dr_2 dr_3$ [m^3] and $d\kappa = d\kappa_1 d\kappa_2 d\kappa_3$ [m^{-3}]. Note that in these
relationships $R_{ij}(\mathbf{r}, t)$ [m^{2}s^{-2}] is considered independent of $\mathbf{x}$ [m], assuming that
turbulence is homogeneous. By setting $\mathbf{r} = 0$ [m] the second integral can give us
the Reynolds stress

$$R_{ij}(0, t) = \langle u_i u_j \rangle = \iiint_{-\infty}^{\infty} \Phi_{ij}(\kappa, t) d\kappa . \qquad (3.38)$$

The velocity spectrum tensor $\Phi_{ij}(\kappa, t)$ [m^{5}s^{-2}] represents the contribution to the
covariance of velocity $\langle u_i u_j \rangle$ [m^{2}s^{-2}] at wavenumber κ [m^{-1}].

3.12   Types of Averaging

So far we have defined the velocity vector mean in turbulent flow as a function of PDF. For instance the mean at any time t [s] was given by

$$\langle U(t) \rangle \equiv \int_{-\infty}^{\infty} V f(V; t) dV . \tag{3.39}$$

However, in turbulent flow an exact PDF is not known or is not possible to measure or calculate. Instead, various types of averaging techniques are performed in simulation and experiments in order to determine an *estimate* for the mean $\langle U(t) \rangle$ [ms^{-1}]. We already introduced the concept of an *ensemble average*, where an identical experiment is repeated N number of times and a velocity component is measured at some location and time N times. The ensemble average is given by

$$\langle U(t) \rangle_N \equiv \frac{1}{N} \sum_{n=1}^{N} U^{(n)}(t) , \tag{3.40}$$

where $U^{(n)}(t)$ is the nth measurement or experiment. Alternatively, and for statistically stationary flow, it is possible to take a *time average* starting at some arbitrary time t [s] for a period of T [s], given by

$$\langle U(t) \rangle_T \equiv \frac{1}{T} \int_{t}^{t+T} U(t') dt' , \tag{3.41}$$

where an integral is used, assuming a continuous measurement of velocity is available. Where a time discrete measurement is available, it is necessary to replace the integral with a sum to calculate the time average. In simulations of homogeneous turbulence, it is also possible to calculate a *spatial average* in a cubic domain of side $\mathcal{L}$ [m] of a component of velocity field $U(\mathbf{x}, t)$ [ms^{-1}], given by

$$\langle U(t) \rangle_{\mathcal{L}} \equiv \frac{1}{\mathcal{L}^3} \iiint_{0}^{\mathcal{L}} U(\mathbf{x}, t) dx_1 \, dx_2 \, dx_3 , \tag{3.42}$$

where an integral is used, assuming a continuous measurement of velocity is available. Where a space discrete measurement is available, it is necessary to replace the integral with a sum to calculate the spatial average. These averages, i.e. $\langle U(t) \rangle_N$, $\langle U(t) \rangle_T$, and $\langle U(t) \rangle_{\mathcal{L}}$ [ms^{-1}], are random variables themselves, but they can be used to *estimate* $\langle U(t) \rangle$ [ms^{-1}], but not to estimate it with certainty.

Problems

3.1 In a one-dimensional mechanical wave the wavelength is $\lambda = 5$ m. What is the value and units of the wave number κ? Calculate the wave number in both units of cycles per unit distance and radians per unit distance.

3.2 Derive the following formula, also known as the Euler's identity, which was known as early as the eighteenth century,

$$e^{i\pi} + 1 = 0 . \tag{3.43}$$

3.3 The autocorrelation function $\rho(s)$ [−] for a flow is given as a function of s in seconds for the range $s = 0$ s to $s = \infty$ s as

$$\rho(s) = \frac{1}{(1+s)^2} . \tag{3.44}$$

Show that the integral timescale of the flow is equal to

$$\overline{\tau} = 1\,\text{s} . \tag{3.45}$$

3.4 The two-point correlation in a flow is given by

$$R_{11}(\mathbf{e}_1 r, \mathbf{x}, t) = f(\mathbf{x}, t)e^{-r/a} , \tag{3.46}$$

where $f(\mathbf{x}, t)$ [m^{2}s^{-2}] is a well-behaved function of time and space that is continuous, differentiable, and integrable, e is the Euler number, and a [m] is a positive real number. If r [m] is the distance in metres, show that for this flow the integral lengthscale in metres can be given by

$$L_{11}(\mathbf{x}, t) = a . \tag{3.47}$$

3.5 A combustion scientist has developed a chamber for combustion experiments, where he premixes air and fuel and then pressurizes the mixture for auto-ignition, i.e. self-ignition without having to use a spark. He is interested in the turbulence properties of combustion at the centre of the chamber in a region of space $2 \times 2 \times 2$ cm^3. He develops a fast camera system to take two-dimensional images on a 2×2 cm^2 plane slicing the centre of this region to obtain images of temperature fluctuations. For each combustion experiment he varies the air and fuel proportions and then takes images at multiple instances of time before and after complete combustion. He then averages temperature fluctuations for each image at a specific time and reports a mean for temperature and a variance for temperature fluctuations as a function of time. (a) Has he assumed the temperature field to be statistically stationary or non-stationary? (b) Has he assumed the temperature field to be statistically homogeneous or non-homogeneous? (c) What type of averaging has he performed, i.e. ensemble, time, or spatial averaging?

References

1. Pope S B (2000) Turbulent flows. Cambridge University Press, Cambridge
2. Barlow R J (1993) Statistics: a guide to the use of statistical methods in the physical sciences, 1st edn. Wiley, New York
3. Walpole R E, Myers, R H, Myers S L et al (2002) Probability & statistics for engineers & scientists, 7th edn. Prentice Hall, Upper Saddle River
4. Kaimal J C, Wyngaard J C, Izumi Y et al (1972) Spectral characteristics of surface-layer turbulence. Q J Roy Meteor Soc 98:563–589
5. Kaimal J C, Wyngaard J C, Haugen D A et al (1976) Turbulence structure in the convective boundary layer. J Atmos Sci 33:2152–2169
6. Aliabadi A A, Moradi M, Clement D et al (2019) Flow and temperature dynamics in an urban canyon under a comprehensive set of wind directions, wind speeds, and thermal stability conditions. Environ Fluid Mech 19:81–109
7. Aliabadi A A, Moradi M, Byerlay R A E (2021) The budgets of turbulence kinetic energy and heat in the urban roughness sublayer. Environ Fluid Mech 21:843–884

Chapter 4
Mean Flow Equations

Abstract This chapter develops the mean flow transport equations for continuity, momentum, and passive scalar, considering turbulence. The Reynolds decomposition is used in the development of the equations, which involves expressing a flow property as the sum of an ensemble average plus the fluctuating component. The concept of mean substantial or total derivative is introduced, which facilitates the notation for the transport equations. The Reynolds stress tensor is defined with components of second-order statistics involving the fluctuations for each component of the velocity vector. The turbulence kinetic energy is defined as one half of the sum of the variances of the three velocity components. The turbulence closure problem is defined. The anisotropy is defined, which is closely related to the gradient-diffusion and turbulent-viscosity hypotheses commonly used for basic turbulence closure schemes. Finally, the turbulent Prandtl and Schmidt numbers are used that relate the turbulent viscosity to the turbulent diffusion coefficients for heat and mass (passive scalar), respectively.

4.1 Overview

The methodology developed in this chapter is adapted from [1]. In Chap. 3 we discussed various statistical quantities that described various aspects of turbulent velocity fields. These included PDF, mean, moments, Reynolds stress, and one-point and two-point correlations. The goal of this chapter is to develop the Navier–Stokes transport equations for various statistical quantities in turbulent flow, as opposed to the equations for the instantaneous quantities. The most basic of these equations are those derived for mean velocity field $\langle \mathbf{U}(\mathbf{x}, t) \rangle$ [ms^{-1}]. As discussed earlier, the velocity field can be decomposed into the mean velocity field and the fluctuating velocity field using *Reynolds decomposition*, i.e.

$$\underbrace{\mathbf{U}(\mathbf{x}, t)}_{\text{Instantaneous Velocity}} = \underbrace{\langle \mathbf{U}(\mathbf{x}, t) \rangle}_{\text{Mean Velocity}} + \underbrace{\mathbf{u}(\mathbf{x}, t)}_{\text{Fluctuating Velocity}} . \quad (4.1)$$

© The Author(s), under exclusive license to Springer Nature Switzerland AG 2022
A. A. Aliabadi, *Turbulence*, Mechanical Engineering Series,
https://doi.org/10.1007/978-3-030-95411-6_4

The continuity equation implies that all $\mathbf{U}(\mathbf{x}, t)$, $\langle \mathbf{U}(\mathbf{x}, t) \rangle$, and $\mathbf{u}(\mathbf{x}, t)$ [ms^{-1}] are solenoidal, given the fact that the mean and differentiation operations commute so that $\langle \nabla.\mathbf{U} \rangle = \nabla.\langle \mathbf{U} \rangle = 0$. This results in

$$\nabla.\langle \mathbf{U}(\mathbf{x}, t) \rangle = 0 , \tag{4.2}$$

$$\nabla.\mathbf{u}(\mathbf{x}, t) = 0 . \tag{4.3}$$

We next focus our attention on the material or substantial derivative to obtain an expression for the material derivative of mean velocity field. By definition of material derivative we can arrive at the expressions [1, 2]

$$\underbrace{\frac{DU_j}{Dt}}_{\text{Material Derivative}} = \underbrace{\frac{\partial U_j}{\partial t}}_{\text{Storage}} + \underbrace{\frac{\partial}{\partial x_i}(U_i U_j)}_{\text{Advection}} , \tag{4.4}$$

$$\left\langle \frac{DU_j}{Dt} \right\rangle = \frac{\partial \langle U_j \rangle}{\partial t} + \frac{\partial}{\partial x_i} \langle U_i U_j \rangle . \tag{4.5}$$

It is then needed to find an expression for $\langle U_i U_j \rangle$ [m^{2}s^{-2}]. This expression can be obtained as follows:

$$\begin{aligned}
\langle U_i U_j \rangle &= \langle (\langle U_i \rangle + u_i)(\langle U_j \rangle + u_j) \rangle \\
&= \langle \langle U_i \rangle \langle U_j \rangle + u_i \langle U_j \rangle + u_j \langle U_i \rangle + u_i u_j \rangle \\
&= \langle U_i \rangle \langle U_j \rangle + \langle u_i u_j \rangle .
\end{aligned} \tag{4.6}$$

Note that the mean for each of the terms $u_i \langle U_j \rangle$ and $u_j \langle U_i \rangle$ [m^{2}s^{-2}] is equal to zero since either u_i or u_j [ms^{-1}] can be given by PDFs centred at zero and symmetric with respect to zero. In other words, the fluctuating velocity field has the same likelihood of being positive or negative and with equal distributions on either side, so that the terms $\langle u_i \langle U_j \rangle \rangle$ and $\langle u_j \langle U_i \rangle \rangle$ [m^{2}s^{-2}] will be zero. With this development the mean of material derivative (Eq. 4.5) for velocity field can be given as

$$\begin{aligned}
\left\langle \frac{DU_j}{Dt} \right\rangle &= \frac{\partial \langle U_j \rangle}{\partial t} + \frac{\partial}{\partial x_i}(\langle U_i \rangle \langle U_j \rangle + \langle u_i u_j \rangle) \\
&= \frac{\partial \langle U_j \rangle}{\partial t} + \langle U_i \rangle \frac{\partial}{\partial x_i} \langle U_j \rangle + \frac{\partial}{\partial x_i} \langle u_i u_j \rangle ,
\end{aligned} \tag{4.7}$$

where the second step follows from $\partial \langle U_i \rangle / \partial x_i = 0$, which is another manifestation of the continuity Eq. 4.2. A useful notation to develop is the *mean substantial derivative* given by

$$\frac{\overline{D}}{\overline{Dt}} \equiv \frac{\partial}{\partial t} + \langle \mathbf{U} \rangle . \nabla \ . \tag{4.8}$$

For any property in the flow, this mean substantial derivative represents its rate of change as the property is following a point moving with the local mean velocity. Using this notation the mean of material or substantial derivative of velocity field can be shown as

$$\underbrace{\left\langle \frac{DU_j}{Dt} \right\rangle}_{\text{Mean of Material Derivative}} = \underbrace{\frac{\overline{D}}{\overline{Dt}} \langle U_j \rangle}_{\text{Mean Substantial Derivative of Mean}} + \underbrace{\frac{\partial}{\partial x_i} \langle u_i u_j \rangle}_{\text{Reynolds Stresses}} \ .$$

$$\tag{4.9}$$

Certainly the mean of material or substantial derivative of velocity field (Eq. 4.9) shall not be confused with, nor is it equal to, the mean substantial derivative (Eq. 4.8). With this consideration, it is possible to develop the equation for the mean momentum such that

$$\underbrace{\frac{\overline{D} \langle U_j \rangle}{\overline{Dt}}}_{\text{Mean Substantial Derivative of Mean}} = \underbrace{\nu \nabla^2 \langle U_j \rangle}_{\text{Surface Forces}} - \underbrace{\frac{\partial \langle u_i u_j \rangle}{\partial x_i}}_{\text{Reynolds Stresses}} - \underbrace{\frac{1}{\rho} \frac{\partial \langle p \rangle}{\partial x_j}}_{\text{Normal and Body Forces}} \ .$$

$$\tag{4.10}$$

The momentum equations for instantaneous and mean velocity field are remarkably similar, but with the crucial difference of the Reynolds stress term. As shall be seen, much of the statistical turbulence modelling theory revolves around parameterizing this Reynolds stress term. This term shows an instance of the *closure problem*, where in turbulent flow equations one always encounters more unknowns than the number of equations available to close the system of equations.

4.2 Tensor Properties

It is worth remembering that Reynolds stresses $\langle u_i u_j \rangle$ [m^{2}s^{-2}] are components of a second-order tensor, with the property that it is symmetric, i.e. $\langle u_i u_j \rangle = \langle u_j u_i \rangle$. The diagonal components of this tensor, i.e. $\langle u_1^2 \rangle = \langle u_1 u_1 \rangle$, $\langle u_2^2 \rangle$, and $\langle u_3^2 \rangle$, are called *normal stresses*, while the off-diagonal components, e.g. $\langle u_1 u_2 \rangle$, are called *shear stresses*. The Reynolds stress tensor can be shown in the matrix as follows:

$$\begin{bmatrix} \langle u_1^2 \rangle & \langle u_1 u_2 \rangle & \langle u_1 u_3 \rangle \\ \langle u_2 u_1 \rangle & \langle u_2^2 \rangle & \langle u_2 u_3 \rangle \\ \langle u_3 u_1 \rangle & \langle u_3 u_2 \rangle & \langle u_3^2 \rangle \end{bmatrix} \ .$$

Given the symmetric property of the Reynolds stress tensor, the matrix is symmetric about the diagonal. The *turbulence kinetic energy* $k(\mathbf{x}, t)$ [m²s⁻²] is defined to be half the trace of the Reynolds stress tensor, i.e.

$$k \equiv \frac{1}{2}\langle \mathbf{u}.\mathbf{u}\rangle = \frac{1}{2}\langle u_i u_i\rangle . \tag{4.11}$$

4.3 Anisotropy

The difference between shear stresses and normal stresses is dependent on the choice of a coordinate system. For instance, if the coordinate system is rotated, the Reynolds stress tensor components may change. It is useful to express the stress tensor as the sum of isotropic and anisotropic parts. This intrinsic distinction implies that the isotropic component of the stresses would never change, regardless of the choice of the reference coordinate system, but the anisotropic part may change. This can be shown as follows [1]:

$$\langle u_i u_j\rangle = \underbrace{\langle u_i u_j\rangle - \frac{2}{3}k\delta_{ij}}_{\text{Anisotropic Part}} + \underbrace{\frac{2}{3}k\delta_{ij}}_{\text{Isotropic Part}} . \tag{4.12}$$

The anisotropic part is also known as *anisotropy* and is notated with the symbol $a_{ij} \equiv \langle u_i u_j\rangle - \frac{2}{3}k\delta_{ij}$ [m²s⁻²]. An important concept is that it is only the anisotropic component that is effective in turbulent transport of momentum. Therefore, the combination of Reynolds stress term and the pressure term in the momentum Eq. 4.10 can be written as

$$\rho\frac{\partial\langle u_i u_j\rangle}{\partial x_i} + \frac{\partial\langle p\rangle}{\partial x_j} = \rho\frac{\partial a_{ij}}{\partial x_i} + \frac{\partial}{\partial x_j}\left(\langle p\rangle + \frac{2}{3}\rho k\right), \tag{4.13}$$

demonstrating that the isotropic component $\frac{2}{3}\rho k$ [kgm⁻¹s⁻²] can be absorbed in the *modified mean pressure*.

4.4 Mean Scalar Equation

The Navier–Stokes transport or conservation Eq. 2.11 for a passive scalar field $\phi(\mathbf{x}, t)$ [−] can be developed further for the mean passive scalar field $\langle\phi(\mathbf{x}, t)\rangle$ [−] as well. Again, the *Reynolds decomposition* can be applied to the passive scalar field

$$\underbrace{\phi(\mathbf{x}, t)}_{\text{Instantaneous Scalar}} = \underbrace{\langle\phi(\mathbf{x}, t)\rangle}_{\text{Mean Scalar}} + \underbrace{\phi'(\mathbf{x}, t)}_{\text{Fluctuating Scalar}} . \quad (4.14)$$

We begin with the conservation Eq. 2.11 for the instantaneous passive scalar field

$$\underbrace{\frac{\partial\phi}{\partial t}}_{\text{Storage}} + \underbrace{\nabla.(\mathbf{U}\phi)}_{\text{Advection}} = \underbrace{\Gamma\nabla^2\phi}_{\text{Diffusion}} . \quad (4.15)$$

The only non-linear term involves the convective flux $\mathbf{U}\phi$ [ms^{-1}], for which we can obtain the following mean:

$$\langle\mathbf{U}\phi\rangle = \langle((\langle\mathbf{U}\rangle + \mathbf{u})(\langle\phi\rangle + \phi'))\rangle$$
$$= \langle\mathbf{U}\rangle\langle\phi\rangle + \langle\mathbf{u}\phi'\rangle . \quad (4.16)$$

The velocity-scalar covariance $\langle\mathbf{u}\phi'\rangle$ [ms^{-1}] is a vector known as *scalar flux*. It quantifies the flow rate of the scalar, per unit time and unit area, due to fluctuating velocity field. This mechanism for transporting the scalar is just as important as the mechanism due to mean velocity field, also known as advection. In fact, in many flows that do not exhibit mean velocity field in some particular direction, the scalar flux is the only mechanism that transports the scalar in that direction. Many boundary layer flows, such as the atmospheric boundary layer, can exhibit this condition under certain circumstances. Taking the mean of the scalar transport Eq. 2.11, we obtain the following results:

$$\frac{\partial\langle\phi\rangle}{\partial t} + \nabla.(\langle\mathbf{U}\rangle\langle\phi\rangle + \langle\mathbf{u}\phi'\rangle) = \Gamma\nabla^2\langle\phi\rangle , \quad (4.17)$$

$$\frac{\overline{D}\langle\phi\rangle}{\overline{Dt}} = \nabla.(\Gamma\nabla\langle\phi\rangle - \langle\mathbf{u}\phi'\rangle) . \quad (4.18)$$

This equation shows another instance of the turbulence *closure problem* where the term $\langle\mathbf{u}\phi'\rangle$ [ms^{-1}] has to be modelled further or parameterized in order to be able to solve the passive scalar equation.

4.5 Gradient-Diffusion and Turbulent-Viscosity Hypotheses

The first attempt to model or parameterize the scalar flux was made by Joseph Valentin Boussinesq (1842–1929). He hypothesized that the mean scalar flux is in the direction and proportional to the negative mean scalar gradient, i.e. $\nabla\langle\phi\rangle$ [m^{-1}]. This is known as the *gradient-diffusion hypothesis* or *turbulent-viscosity hypothesis*.

The constant of proportionality is the *turbulent diffusivity* that is itself a function of space and time, $\Gamma_T(\mathbf{x}, t)$ [m^{2}s^{-1}], i.e. [1]

$$\langle \mathbf{u}\phi' \rangle = -\Gamma_T \nabla \langle \phi \rangle \ . \tag{4.19}$$

The subscript T is for turbulence, and Γ_T [m^{2}s^{-1}] is *turbulent diffusivity* and should not be confused with *molecular diffusivity* Γ [m^{2}s^{-1}]. This hypothesis is similar to Fourier's law of heat conduction and Fick's law of molecular diffusion, where the heat or scalar flux is always in the direction from a hot or high concentration region towards a cold or low concentration region. In other words, the *diffusion* is in the negative direction of the gradient. Because of this, another term for this hypothesis is the *down-gradient hypothesis*. It is remarkable that, at least conceptually, the scalar turbulent flux is formulated in the similar manner as the molecular flux. Due to this reason, it is possible to combine the molecular and turbulent diffusivities to arrive at an *effective diffusivity* given by

$$\underbrace{\Gamma_{\text{eff}}(\mathbf{x}, t)}_{\text{Effective Diffusivity}} = \underbrace{\Gamma}_{\text{Molecular Diffusivity}} + \underbrace{\Gamma_T(\mathbf{x}, t)}_{\text{Turbulent Diffusivity}} \ . \tag{4.20}$$

The effective diffusivity may be referred to as effective heat diffusivity associated with a temperature or effective mass diffusivity associated with a passive scalar concentration. With this consideration, it is possible to express the mean scalar conservation Eq. 4.18 as

$$\underbrace{\frac{\overline{D}\langle \phi \rangle}{\overline{D}t}}_{\text{Mean Substantial Derivative of Mean}} = \underbrace{\nabla.(\Gamma_{\text{eff}}\nabla \langle \phi \rangle)}_{\text{Diffusion of Mean}} \ . \tag{4.21}$$

For the mean momentum transport Eq. 4.10, the gradient-diffusion hypothesis is more delicate to implement. In this case we should relate the anisotropic part of the Reynolds stress $\langle u_i u_j \rangle - \frac{2}{3}k\delta_{ij}$ [m^{2}s^{-2}] (Eq. 4.12) to the mean rate of strain (Eq. 2.17) by

$$\langle u_i u_j \rangle - \frac{2}{3}k\delta_{ij} = -v_T \left(\frac{\partial \langle U_i \rangle}{\partial x_j} + \frac{\partial \langle U_j \rangle}{\partial x_i} \right)$$

$$= -2v_T \overline{S}_{ij} \ , \tag{4.22}$$

where the positive scalar coefficient v_T [m^{2}s^{-1}] is the momentum diffusivity also known as *turbulent viscosity* or *eddy viscosity*. With these considerations, the mean momentum Eq. 4.10 becomes

$$\underbrace{\frac{\overline{D}}{Dt}\langle U_j\rangle}_{\text{Mean Substantial Derivative of Mean}} = \underbrace{\frac{\partial}{\partial x_i}\left[\nu_{\text{eff}}\left(\frac{\partial\langle U_i\rangle}{\partial x_j}+\frac{\partial\langle U_j\rangle}{\partial x_i}\right)\right]}_{\text{Surface Forces and Reynolds Stress}} - \underbrace{\frac{1}{\rho}\frac{\partial}{\partial x_j}\left(\langle p\rangle+\frac{2}{3}\rho k\right)}_{\text{Modified Pressure}},$$

$$\tag{4.23}$$

$$\underbrace{\nu_{\text{eff}}(\mathbf{x},t)}_{\text{Effective Viscosity}} = \underbrace{\nu}_{\text{Molecular Viscosity}} + \underbrace{\nu_T(\mathbf{x},t)}_{\text{Turbulent Viscosity}}, \tag{4.24}$$

where ν_{eff} [m^{2}s^{-1}] is the *effective viscosity* and $\langle p\rangle + \frac{2}{3}\rho k$ [kgm^{-1}s^{-2}] is the modified pressure. It must be recalled that the gradient-diffusion hypothesis (Eq. 4.22) can only be applied to the anisotropic part of the Reynolds stress because the momentum is transported only by the anisotropic part of the Reynolds stress. This consideration requires the modified pressure instead of pressure to remain in the transport equation.

Similar to molecular diffusivity, turbulent diffusivity can be non-dimensionalized as well. The turbulent Prandtl number gives the ratio of momentum diffusivity to heat diffusivity by

$$Pr_T = \frac{\nu_T}{\Gamma_T}. \tag{4.25}$$

Likewise, the turbulent Schmidt number gives the ratio of momentum diffusivity to mass diffusivity by

$$Sc_T = \frac{\nu_T}{\Gamma_T}. \tag{4.26}$$

The turbulent Prandtl and Schmidt numbers are not material constants and may vary greatly according to flow conditions. In addition, they may not even be equal to each other. However, under special cases, for instance for air, where the Lewis number $Le = Sc/Pr \approx 1$ [−], we can employ the concept of *Reynolds analogy* and assume $Pr = Sc$ [−] and $Pr_T = Sc_T$ [−] [3, 4].

It is worth noting that the gradient-diffusion hypothesis (Eq. 4.19) is a simple hypothesis that enables closing a turbulence model equation. However, this hypothesis is not valid for many flows, even the simplest flows, and can lead to significant errors in the calculation. So it must be used with extreme care. It is only in light of historical developments and its limited use that we introduce this hypothesis as a closure scheme for a turbulence model equation. Furthermore, the turbulent diffusivity Γ_T [m^{2}s^{-1}] itself needs to be modelled accurately in order to arrive at more accurate closure schemes. Again, much of the theory of turbulence modelling is focused around the parameterization for the turbulent diffusivity, of course for gradient-diffusion models. These topics will be revisited in Chap. 15.

Problems

4.1 The simplest model for the turbulent Prandtl number assumes $Pr_T = 1$ [−]. However, numerous experimental observations suggest that Pr_T [−] can be variable even for the same fluid under different flow circumstances. For instance, flows near walls exhibit a different turbulent Prandtl number compared to flows in the internal region of a fluid domain away from the walls [3]. Assuming $Pr_T = 0.85$ [−], express a relationship between momentum diffusivity ν_T [m²s⁻¹] and heat diffusivity Γ_T [m²s⁻¹].

4.2 Using the gradient-diffusion hypothesis, the surface and Reynolds stress forces in the momentum equation are expressed by the term

$$\frac{\partial}{\partial x_i}\left[\nu_{\text{eff}}\left(\frac{\partial \langle U_i\rangle}{\partial x_j} + \frac{\partial \langle U_j\rangle}{\partial x_i}\right)\right]. \tag{4.27}$$

Reason why this term cannot be expressed and is not equal to the following term under general conditions:

$$\nu_{\text{eff}}\frac{\partial}{\partial x_i}\left[\left(\frac{\partial \langle U_i\rangle}{\partial x_j} + \frac{\partial \langle U_j\rangle}{\partial x_i}\right)\right]. \tag{4.28}$$

4.3 Suppose that in a flow the mean passive scalar field $\langle \phi\rangle$ [−] can be expressed as a function of x, y, and z coordinates along the Cartesian system with unit vectors **i**, **j**, and **k**, respectively, such that

$$\langle \phi\rangle = \phi_0(x + y^2 - z^3), \tag{4.29}$$

where ϕ_0 [−] is a constant. Assuming that the gradient-diffusion hypothesis applies to this flow and passive scalar field, show that the passive scalar flux $\langle \mathbf{u}\phi'\rangle$ [ms⁻¹] is along the following unit vector at location $(x, y, z) = (1, 1, 1)$, i.e.

$$\frac{\langle \mathbf{u}\phi'\rangle}{|\langle \mathbf{u}\phi'\rangle|}\Big|_{(1,1,1)} = -\frac{1}{\sqrt{14}}\mathbf{i} - \frac{2}{\sqrt{14}}\mathbf{j} + \frac{3}{\sqrt{14}}\mathbf{k}. \tag{4.30}$$

4.4 Given a constant effective viscosity ν_{eff} [m²s⁻¹] that accounts for effects of molecular and turbulent diffusion, we wish to develop a one-dimensional (1D) momentum transport equation. Suppose we use the Cartesian coordinate system with coordinate axes of x, y, and z, and velocities corresponding to these axes being $U = \langle U\rangle + u$, $V = \langle V\rangle + v$, and $W = \langle W\rangle + w$ [ms⁻¹], respectively. Further, we assume that mean flow is only in the x direction parallel to the surface and that the direction z is normal to the surface, i.e. $\langle V\rangle = \langle W\rangle = 0$ ms⁻¹. We assume steady state conditions. In addition, we assume that the mean velocity $\langle U\rangle$ [ms⁻¹] in the x and y directions does not change. Also we assume that the modified pressure has a constant gradient in the x direction. Show that the 1D momentum equation then simplifies to

$$0 = \nu_{\text{eff}}\frac{d^2\langle U\rangle}{dz^2} - \tau, \tag{4.31}$$

where τ is the constant modified pressure gradient in the x direction divided by density. Demonstrate why the storage and advection terms in the momentum transport equation have vanished.

4.5 Given a non-constant turbulent viscosity ν_T [m^{2}s^{-1}] that accounts for effects of molecular and turbulent diffusion and a turbulent Prandtl number $Pr_T = 1$ [−], we wish to develop a one-dimensional (1D) heat (temperature) transport equation. Suppose we use the Cartesian coordinate system with coordinate axes of x, y, and z, and velocities corresponding to these axes being $U = \langle U \rangle + u$, $V = \langle V \rangle + v$, and $W = \langle W \rangle + w$ [ms^{-1}], respectively. Further, we assume that mean flow is only in the x direction parallel to the surface and that the direction z is normal to the surface, i.e. $\langle V \rangle = \langle W \rangle = 0$ ms^{-1}. We assume steady state conditions. In addition, we assume that the mean velocity $\langle U \rangle$ [ms^{-1}] in the x and y directions does not change. We also assume a constant heat sink or source for temperature by a uniform rate of cooling or heating in the domain. Show that the 1D heat equation then simplifies to

$$0 = \frac{\partial}{\partial z}\left(\nu_T \frac{\partial \langle T \rangle}{\partial z}\right) - \gamma , \qquad (4.32)$$

where the term γ [Ks^{-1}] is a sink or source for temperature by uniform rate of cooling or heating in the domain. Demonstrate why the storage and advection terms in the heat transport equation have vanished.

4.6 Given a non-constant turbulent viscosity ν_T [m^{2}s^{-1}] that accounts for effects of molecular and turbulent diffusion and a turbulent Schmidt number $Sc_T = 1$ [−], we wish to develop a transient one-dimensional (1D) passive scalar transport equation. Suppose we use the Cartesian coordinate system with coordinate axes of x, y, and z and velocities corresponding to these axes being $U = \langle U \rangle + u$, $V = \langle V \rangle + v$, and $W = \langle W \rangle + w$ [ms^{-1}], respectively. Further, we assume that mean flow is only in the x direction parallel to the surface and that the direction z is normal to the surface, i.e. $\langle V \rangle = \langle W \rangle = 0$ ms^{-1}. In addition, we assume that the mean velocity $\langle U \rangle$ in the x and y directions does not change. Show that the 1D passive scalar transport equation then simplifies to

$$\frac{\partial \langle \phi \rangle}{\partial t} = \frac{\partial}{\partial z}\left(\nu_T \frac{\partial \langle \phi \rangle}{\partial z}\right) . \qquad (4.33)$$

Demonstrate why the advection term in the passive scalar transport equation has vanished.

4.7 The turbulence intensity, also often referred to as turbulence level, is defined as

$$I \equiv \frac{\mathbf{u}}{\langle \mathbf{U} \rangle} , \qquad (4.34)$$

where $\mathbf{u}$ [ms^{-1}] is the root mean square of the turbulent velocity fluctuations and $\langle \mathbf{U} \rangle$ [ms^{-1}] is the mean velocity. Assuming that turbulence is isotropic, show that

$$I \equiv \frac{\sqrt{\frac{2}{3}k}}{\langle U \rangle} \, , \tag{4.35}$$

where k [m^{2}s^{-2}] is the turbulence kinetic energy. For atmospheric flows I [$-$] is typically less than one, and under gusty conditions, it can be equal to or larger than one.

4.8 A meteorologist is working on the following problem. The atmospheric boundary layer can be approximated by a one-dimensional transport equation for heat, i.e. temperature, in the vertical direction (z) over short distances, where atmospheric pressure can be assumed constant with no temperature lapse rate. The one-dimensional assumption is based on horizontal homogeneity of the atmosphere and no mean vertical motion of air. Using this approximation, the vertical transport of heat or temperature can be expressed as

$$\frac{\partial \langle T \rangle}{\partial t} = -\frac{\partial \langle tw \rangle}{\partial z} \, , \tag{4.36}$$

where $T = \langle T \rangle + t$ [K] is the temperature expressed as the sum of mean and turbulent fluctuating parts, and $W = \langle W \rangle + w$ [ms^{-1}] is the wind speed in the vertical direction expressed as the sum of mean and turbulent fluctuating parts. Of course it is assumed that $\langle W \rangle = 0$ ms^{-1}. Assume that the turbulent heat flux $\langle tw \rangle$ [Kms^{-1}] in a range of heights from 10 to 30 m for a diurnal time is given as

$$\langle tw \rangle = -\frac{1}{1+z} \, , \tag{4.37}$$

where z is the height in [m]. Help the meteorologist calculate how much the temperature of the atmosphere changes at a height of $z = 15$ m over a time duration of 100 s. Is the atmosphere cooling or heating in this scenario?

4.9 A meteorologist is deriving the energy equation (i.e. temperature equation) near the earth surface under steady state conditions. The atmospheric boundary layer can be approximated by a one-dimensional transport equation for heat, i.e. temperature, in the vertical direction (z) over short distances, where atmospheric pressure can be assumed constant with no temperature lapse rate. The one-dimensional assumption is based on horizontal homogeneity of the atmosphere and no mean vertical motion of air. (a) Using this approximation, show that the temperature near the surface can be expressed as

$$0 = \frac{d}{dz} \left(\Gamma \frac{d \langle T \rangle}{dz} - \langle tw \rangle \right) \, , \tag{4.38}$$

where $T = \langle T \rangle + t$ [K] is the temperature expressed as the sum of mean and turbulent fluctuating parts, $W = \langle W \rangle + w$ [ms^{-1}] is the wind speed in the vertical direction expressed as the sum of mean and turbulent fluctuating parts, and Γ [m^{2}s^{-1}] is the thermal diffusivity. (b) Using the turbulent-viscosity assumption, show that this equation can be written as

$$0 = \frac{d}{dz}\left[\left(\frac{\nu}{Pr} + \frac{\nu_T}{Pr_T}\right)\frac{d\langle T\rangle}{dz}\right]. \tag{4.39}$$

4.10 Consider a two-dimensional flow field in which velocity can be given by time averaging along x and z components using Reynolds averaging such that $U = \langle U\rangle + u$ and $W = \langle W\rangle + w$ [ms^{-1}]. Consider that there is also a passive scalar that is being transported by the velocity field and can be shown using Reynolds averaging by $S = \langle S\rangle + s$ [−]. Of course the velocity and passive scalar are functions of space and time. Suppose that one wishes to calculate the total flux of the passive scalar along the x and z directions at a given point. Show that the total fluxes can be calculated using the following expressions:

$$\underbrace{\langle SU\rangle}_{\text{Total flux}} = \underbrace{\langle S\rangle\langle U\rangle}_{\text{Advective flux}} + \underbrace{\langle su\rangle}_{\text{Turbulent flux}}, \tag{4.40}$$

$$\underbrace{\langle SW\rangle}_{\text{Total flux}} = \underbrace{\langle S\rangle\langle W\rangle}_{\text{Advective flux}} + \underbrace{\langle sw\rangle}_{\text{Turbulent flux}}. \tag{4.41}$$

In another word, the total flux can be calculated as the sum of advective flux and turbulent flux.

4.11 A meteorologist is developing a model for vertical transport of atmospheric species within the atmospheric boundary layer. The boundary layer is a two-dimensional flow field in which velocity can be given by time averaging along x and z components using Reynolds averaging such that $U = \langle U\rangle + u$ and $W = \langle W\rangle + w$ [ms^{-1}]. Here x is along wind direction, or parallel to earth surface, and z is normal to earth surface positive upward. Consider that there is also a passive scalar that is being released at the surface and transported upward due to molecular and turbulent diffusion transport mechanisms. Using Reynolds averaging, the passive scalar can be shown by $S = \langle S\rangle + s$ [kgm^{-3}]. The vertical flux of the species along the z direction can be given by

$$\underbrace{F_S}_{\text{Vertical flux of } s} = \underbrace{-\frac{\nu}{Sc}\frac{d\langle S\rangle}{dz}}_{\text{Molecular diffusion flux}} + \underbrace{\langle sw\rangle}_{\text{Turbulent diffusion flux}}, \tag{4.42}$$

where $\nu = 1.5 \times 10^{-5}$ m^{2}s^{-1} is the molecular viscosity of air and $Sc = 1$ [−] is the molecular Schmidt number. Here, $\langle W\rangle = 0$ ms^{-1}; therefore, there is no advective flux. Consider that the mean passive scalar varies exponentially away from the surface, i.e. $\langle S\rangle = e^{-z}$ [kgm^{-3}]. (a) Perform unit analysis to find the appropriate unit for F_S. (b) Use the turbulent diffusion hypothesis to express the flux only in terms of v, Sc, v_T, Sc_T, and $\langle S \rangle$, where subscript T denotes turbulent. (c) Now consider that at $z = 1$ m, $v_T = 5$ m^{2}s^{-1} and $Sc_T = 1$ [$-$]. Calculate F_S. (d) In the previous calculation, what fraction of the flux is due to molecular diffusion and what fraction of the flux is due to turbulent diffusion? What do you conclude?

References

1. Pope S B (2000) Turbulent flows. Cambridge University Press, Cambridge
2. Acheson D J (1990) Elementary fluid dynamics. Oxford University Press, Oxford
3. Reynolds A J (1975) The prediction of turbulent Prandtl and Schmidt numbers. Int J Heat Mass Transfer 18:1055–1069
4. Flores F, Gerreaud R, Muñoz R C (2013) CFD simulations of turbulent buoyant atmospheric flows over complex geometry: solver development in OpenFOAM. Comput Fluids 82:1–13

Chapter 5
Wall Flows

Abstract This chapter introduces wall flows, which are flows that are bounded by at least one wall. The transport equation for momentum near a wall is developed. The concept of boundary layer is defined. Viscous and turbulent shear stresses near a wall are characterized. The inner and outer layers near a wall are defined, which in turn are subdivided into sublayers mainly comprising of viscous, buffer, and log-law layers. The law of the wall is formulated for both momentum and temperature.

5.1 Overview

Most turbulent flows are bounded by at least in part one or more solid surfaces. For instance flows in pipes or buildings are fully surrounded by surfaces and are the so-called *internal flows*. Flows over the earth surface, around a ship hull, and over airfoil of an aircraft or vehicle are partially surrounded by surfaces and are the so-called *external flows*. Understanding flows near surfaces or walls is of fundamental importance in turbulence studies.

5.2 Transport Equations and the Balance of Mean Forces

We do not restrict our discussion to internal or external flows but will discuss *fully developed flows* near walls, defined as a state where velocity statistics no longer vary in the direction of the flow. For instance, this is the case in short range for atmospheric flows near the earth surface on flat terrain [1, 2]. Suppose the flow near the wall is in the x direction, the direction normal to the wall is y, and the spanwise direction is z, in the Cartesian coordinate system. The corresponding mean velocities are $\langle U \rangle$, $\langle V \rangle$, and $\langle W \rangle$ [ms^{-1}], respectively.

As shown in Fig. 5.1, suppose that the influence of the wall on the flow is diminished at a distance of δ [m] away from the wall in the normal direction, where the free stream velocity is almost achieved completely, say by 99%, $0.99U_0 = \langle U \rangle_{y=\delta}$ [ms^{-1}]. The distance δ [m] represents the thickness of the *boundary layer*

© The Author(s), under exclusive license to Springer Nature Switzerland AG 2022
A. A. Aliabadi, *Turbulence*, Mechanical Engineering Series,
https://doi.org/10.1007/978-3-030-95411-6_5

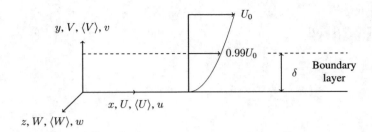

Fig. 5.1 Schematic of two-dimensional wall flow and boundary layer for the fully developed flow near the wall

that is a thin layer of fluid close to the solid surface of a wall in contact with a moving stream, in which the flow velocity varies from zero at the wall up to close to the value of the free stream velocity at the boundary (usually 99% of the free stream velocity). δ [m] is also known as *boundary layer thickness*. The bulk velocity of the flow within the boundary layer is given by

$$\overline{U} \equiv \frac{1}{\delta} \int_0^{\delta} \langle U \rangle dy \ . \tag{5.1}$$

Since $\langle U \rangle$ [ms^{-1}] does not vary as a function of x for fully developed flow and that $\langle W \rangle = 0$ ms^{-1}, the continuity Eq. 2.2 reduces to

$$\frac{d\langle V \rangle}{dy} = 0 \ . \tag{5.2}$$

Note that the continuity equation is written as a derivative as opposed to partial derivative. This is the case since all mean variables in the fully developed region are only a function of y [m]. The boundary condition at the wall is the *no-slip boundary condition* implying that the fluid does not move near a stationary wall, which implies that $\langle V \rangle = 0$ ms^{-1} everywhere. This implies that $V = v$ [ms^{-1}], i.e. the velocity in the y direction is equal to the turbulent fluctuation in that direction. With this development, the mean momentum Eq. 4.10 in the y direction can be written as

$$0 = - \underbrace{\frac{d}{dy}\langle v^2 \rangle}_{\text{Reynolds Stress}} - \underbrace{\frac{1}{\rho}\frac{\partial \langle p \rangle}{\partial y}}_{\text{Normal and Body Forces}} \ . \tag{5.3}$$

Again note that in the Reynolds stress term we have a derivative, but in the normal and body forces term we have a partial derivative. This equation can be integrated from $y = 0$ to an arbitrary height. At the wall $\langle v^2 \rangle_{y=0} = 0$ m^{2}s^{-2} but $\langle p(x) \rangle_{y=0} = P_w(x)$ [kgm^{-1}s^{-2}]; therefore, the integration of the mean momentum Eq. 5.3 in the y direction can be written as

$$\langle v^2 \rangle + \langle p \rangle/\rho = p_w(x)/\rho \ . \tag{5.4}$$

If we now take the derivative of this equation with respect to x, we find that the partial derivative of the mean pressure at any height is equal to the derivative of pressure at the wall in the x direction, i.e.

$$\frac{\partial \langle p \rangle}{\partial x} = \frac{dp_w(x)}{dx} \, , \tag{5.5}$$

which states that the mean axial pressure drop is uniform across the flow, no matter how close or far a location is from the wall. We can now write the momentum Eq. 4.10 in the x direction as

$$0 = \underbrace{\nu \frac{d^2 \langle U \rangle}{dy^2}}_{\text{Surface Forces}} - \underbrace{\frac{d}{dy} \langle uv \rangle}_{\text{Reynolds Stress}} - \underbrace{\frac{1}{\rho} \frac{\partial \langle p \rangle}{\partial x}}_{\text{Normal and Body Forces}} . \tag{5.6}$$

If we group the first two terms on the right hand side, this equation can be written as

$$\frac{d\tau}{dy} = \frac{dp_w(x)}{dx} \, , \tag{5.7}$$

where τ [kgm^{-1}s^{-2}] is the *total shear stress* accounting for surface forces and the Reynolds stress, but not the normal and body forces, which are accounted for in the pressure term. The total shear stress can be written as

$$\underbrace{\tau}_{\text{Total stress}} = \underbrace{\rho \nu \frac{d \langle U \rangle}{dy}}_{\text{Viscous stress}} \underbrace{-\rho \langle uv \rangle}_{\text{Turbulent stress}} . \tag{5.8}$$

Since there is no net acceleration in this flow, in Eq. 5.7 the axial normal stress gradient $dp_w(x)/dx$ [kgm^{-2}s^{-2}] is balanced by the cross-stream shear stress gradient $d\tau/dy$ [kgm^{-2}s^{-2}]. The differential Eq. 5.7 written for this balance states that the shear stress is only a function of y [m], while the normal stress is only a function of x [m]. This implies that both $dp_w(x)/dx$ and $d\tau/dy$ [kgm^{-2}s^{-2}] must be constants. The value of τ [kgm^{-1}s^{-2}] at the wall, i.e. τ_w [kgm^{-1}s^{-2}], is known as *wall shear stress*

$$\tau_w = \tau(y = 0) \, . \tag{5.9}$$

Figure 5.2 shows the total shear stress, composed of viscous and turbulent stresses, as a function of distance from the wall. It can be seen that the viscous stress dominates regions near the wall, while the turbulent stress is enhanced at distances away from the wall. It can also be noted that the total stress gradually declines below

Fig. 5.2 Total shear stress for fluid flow near a wall, comprised of viscous and turbulent stresses, as a function of distance from the wall

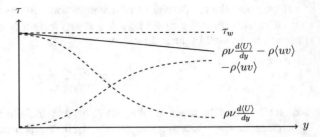

the wall shear stress with increasing distance from the wall. This is expected since the wall friction on the fluid should be the greatest at the vicinity of the wall.

The wall shear stress can be normalized by a reference velocity to give the *skin-drag coefficient*. This type of skin drag is in addition to the pressure drag as is described by the *pressure-drag coefficient*, which is caused by pressure differential across a bluff body in flow. In fact skin drag imposes a type of drag on the fluid in addition to pressure drag. The skin-drag and pressure-drag coefficients can be given as

$$c_f \equiv \frac{\tau_w}{\frac{1}{2}\rho U_0^2} , \tag{5.10}$$

$$c_p \equiv \frac{\Delta p}{\frac{1}{2}\rho U_0^2} , \tag{5.11}$$

where U_0 [ms^{-1}] is the free stream velocity. Most objects immersed in flow impose both types of drag on the fluid: skin drag and pressure drag. Compared to laminar flow, skin drag associated with turbulent flow near walls is higher, so more friction on the fluid may be imposed. However, turbulent flow near parts of a bluff body may reduce the pressure drag on the fluid appreciably, so that the net drag on the fluid (and thus the body using Newton's third law of motion) may be reduced. This knowledge is used in designing golf balls with dimples to create turbulence and ultimately lower the net drag coefficient of the ball by substantially reducing the pressure drag (even at the cost of modestly increasing the skin drag).

5.3   The Shear Stress Near Wall

The total shear stress (Eq. 5.8) at any distance away from the wall, i.e. $\tau(y)$ [kgm^{-1}s^{-2}], is the sum of the viscous stress $\rho v d\langle U\rangle/dy$ and the Reynolds stress $-\rho\langle uv\rangle$ [kgm^{-1}s^{-2}]. We know that at the wall $\mathbf{U}(\mathbf{x}, t) = 0$ ms^{-1}, and therefore, there cannot be any turbulent fluctuations and consequently no Reynolds stress.

Consequently, the wall shear stress is only caused by surface forces, i.e.

$$\tau_w \equiv \rho v \left(\frac{d\langle U \rangle}{dy} \right) . \tag{5.12}$$

The kinematic viscosity v [m^{2}s^{-1}], density ρ [kgm^{-3}], and shear stress near the wall τ_w [kgm^{-1}s^{-2}] are important parameters in turbulence studies. After all, they determine the skin friction and other flow properties. From these quantities we can define *viscous scales* that are appropriate velocity scales and lengthscales associated with the near-wall region. The *friction velocity* and *viscous lengthscale* are defined as

$$u_\tau \equiv \sqrt{\frac{\tau_w}{\rho}} , \tag{5.13}$$

$$\delta_v \equiv v \sqrt{\frac{\rho}{\tau_w}} = \frac{v}{u_\tau} . \tag{5.14}$$

The *friction Reynolds number* is defined with friction velocity and the entire height of the boundary layer, i.e.

$$Re_\tau \equiv \frac{u_\tau \delta}{v} = \frac{\delta}{\delta_v} . \tag{5.15}$$

An important parameter in the study of turbulence near walls is the distance from the wall measured in viscous lengths (Eq. 5.14), or *wall units*, denoted by

$$y^+ \equiv \frac{y}{\delta_v} = \frac{u_\tau y}{v} . \tag{5.16}$$

This quantity is similar to a local Reynolds number, so that its magnitude can be expected to determine the relative importance of viscous and turbulent processes. If y^+ [−] is small, then the relative importance of viscous processes is higher, while if y^+ [−] is large, the relative importance of turbulent processes is higher. In fact, knowledge of the magnitude of y^+ [−] is sufficient to understand which wall regime is present with respect to viscous and turbulent processes. Figure 5.3 shows the classification of wall layers and regions (sublayers). In general, we have two layers: the *inner layer* and the *outer layer* that of course overlap significantly.

Another important parameter in the study of turbulence near walls is the non-dimensional mean velocity, which is the mean velocity divided by the friction velocity, defined by

$$u^+ \equiv \frac{\langle U \rangle}{u_\tau} . \tag{5.17}$$

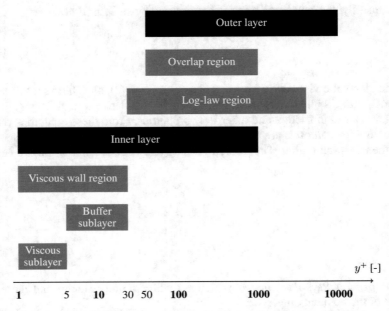

Fig. 5.3 Wall layers (black-filled) and regions (grey-filled) defined in terms of y^+ [−]

A critical task in studies of turbulence near walls is to find a functional relationship between y^+ [−] and u^+ [−], also known as the *law of the wall*, i.e.

$$u^+ = f_w(y^+) . \tag{5.18}$$

Of course there is no universal law of the wall since the law of the wall can change according to the physics of the flow. However, for typical flows, the law of the wall yields three distinct wall regions or sublayers: *viscous sublayer*, *buffer sublayer*, and *log-law sublayer*.

5.4 Viscous, Buffer, and Log-Law Sublayers

The viscous, buffer, and log-law sublayers were first studied by Theodore von Kármán (1881–1963) [3]. In the *viscous sublayer* $y^+ < 5$ [−], the functional relationship between y^+ [−] and u^+ [−] is linear, so that it can be given as

$$u^+ = y^+ . \tag{5.19}$$

In the *buffer sublayer* $5 < y^+ < 30$, there is no simple relationship between y^+ [−] and u^+ [−]. This is the region where both viscous and turbulent processes can

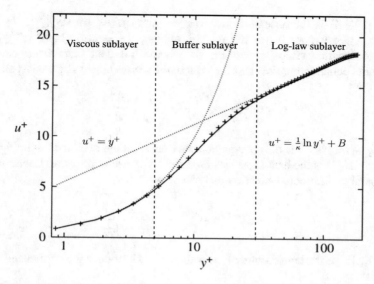

Fig. 5.4 Law of the wall and classifications of viscous, buffer, and log-law sublayers

be present. In the *log-law sublayer* $y^+ > 30$ [−], the relationship between y^+ [−] and u^+ [−] is logarithmic such that

$$u^+ = \frac{1}{\kappa} \ln y^+ + B \,, \tag{5.20}$$

$$\kappa = 0.41, B = 5.2 \,, \tag{5.21}$$

where κ [−] is the von Kármán constant and B [−] is another constant. These constants can be slightly different depending on the physics of the flow. The *law of the wall* for the viscous, buffer, and log-law sublayers is demonstrated in Fig. 5.4.

5.5 Law of the Wall for Temperature

It is possible to develop a law of the wall for the energy (temperature) equation. Consider a flow along the x axis near a wall with axis y normal to the wall. Assuming horizontal homogeneity and using the turbulent-viscosity assumption, the energy equation can be written as

$$0 = \frac{d}{dy} \left[\left(\frac{\mu}{Pr} + \frac{\mu_T}{Pr_T} \right) \frac{d\langle T \rangle}{dy} \right] \,, \tag{5.22}$$

where $\langle T \rangle$ [K] is the ensemble mean temperature, $\mu = \rho \nu$ [kgm^{-1}s^{-1}] is the dynamic viscosity, Pr [$-$] is the Prandtl number, $\mu_T = \rho \nu_T$ [kgm^{-1}s^{-1}] is the turbulent dynamic viscosity, and Pr_T [$-$] is the turbulent Prandtl number. This differential equation requires that the term in the square bracket be a constant, i.e.

$$C = -\frac{q_w}{C_p} = \left(\frac{\mu}{Pr} + \frac{\mu_T}{Pr_T} \right) \frac{d\langle T \rangle}{dy} , \tag{5.23}$$

where q_w [Jm^{-2}s^{-1}] is the specific heat flux between the wall and the fluid in and C_p [Jkg^{-1}K^{-1}] is the heat capacity of the fluid at constant pressure. Equation 5.23 is a separable differential equation that can be written as

$$\int_{T_w}^{\langle T \rangle} d\langle T \rangle = -\frac{q_w}{C_p} \int_0^y \frac{dy}{\frac{\mu}{Pr} + \frac{\mu_T}{Pr_T}} , \tag{5.24}$$

where T_w [K] is the temperature of the wall. We can introduce the normalized values such that

$$y^+ \equiv \frac{y u_\tau}{\nu} , \tag{5.25}$$

$$T^+ \equiv \frac{(T_w - \langle T \rangle)\rho C_p u_\tau}{q_w} , \tag{5.26}$$

where ν [m^{2}s^{-1}] is the kinematic viscosity, ρ [kgm^{-3}] is the fluid density, and u_τ [ms^{-1}] is the friction velocity. Using the relationships above we can derive the following equation for T^+ [$-$]:

$$T^+ = \int_0^{y^+} \frac{dy^+}{\frac{1}{Pr} + \frac{(\nu_T/\nu)}{Pr_T}} , \tag{5.27}$$

where ν_T [m^{2}s^{-1}] is the turbulent viscosity. This integral is not useful practically. However, we can show that if two layers are considered, one laminar where $y^+ = 0 \rightarrow y_{crit}^+$ [$-$] and one turbulent where $y^+ > y_{crit}^+$ [$-$], T^+ [$-$] can be written as

$$T^+ = \int_0^{y_{crit}^+} Pr dy^+ + \int_{y_{crit}^+}^{y^+} \frac{dy^+}{\frac{(\nu_T/\nu)}{Pr_T}} . \tag{5.28}$$

Using the mixing length theory, to be fully discussed in Chap. 15, we can relate turbulent viscosity and distance away from the wall by $\nu_T = u_\tau \kappa y$ [m^{2}s^{-1}] (Eq. 15.11), where $\kappa = 0.41$ [$-$] is the von Kármán constant. We can show that in such a case the T^+ [−] equation can be written as [4]

$$T^+ = y^+_{crit} Pr + \frac{Pr_T}{\kappa} \ln\left(\frac{y^+}{y^+_{crit}}\right) , \qquad (5.29)$$

which is known as the law of the wall for the temperature field. For instance, for air, if $Pr = 0.7$ [−], $Pr_T = 0.85$ [−], and $y^+_{crit} = 13.2$ [−], then the following expression can be obtained as the law of the wall if y^+ [−] is in the turbulent layer [5]

$$T^+ = 2.075 \ln(y^+) + 3.9 . \qquad (5.30)$$

Problems

5.1 Show that the Reynolds number based on the viscous scales near a wall is identically unity, i.e.

$$\frac{u_\tau \delta_v}{\nu} = 1 . \qquad (5.31)$$

5.2 The law of the wall states that for the viscous sublayer $u^+ = y^+$ [−] and for the log-law sublayer $u^+ = \frac{1}{\kappa} \ln y^+ + B$ [−], where $\kappa = 0.41$ [−] and $B = 5.2$ [−]. It is desired to calculate the point of intersection of the two equations, i.e. the value of y^+ [−] for which the same value of u^+ [−] is obtained using either equation. An exact calculation of such y^+ [−] requires solving a non-linear equation involving both y^+ [−] and $\ln y^+$ [−]. However, a first approximation for the suitable y^+ can be obtained by replacing the log-law equation with its first-order Taylor expansion expressed about $y^+ = 10$ [−]. Doing so, the non-linear equation can be linearized and solved conveniently. Using this approach, show that the value of y^+ [−] in question is approximately

$$y^+ \approx 11.1 . \qquad (5.32)$$

5.3 Show that in general

$$c_f = 2\left(\frac{u_\tau}{U_0}\right)^2 . \qquad (5.33)$$

5.4 Various thickness quantities can be defined to characterize wall flows. These thickness quantities themselves can be functions of distance x [m] downstream of the flow. In this chapter, the *boundary layer thickness* was defined and noted with $\delta(x)$ [m]. The *displacement thickness* can be defined as

$$\delta^*(x) \equiv \int_0^\infty \left(1 - \frac{\langle U \rangle(y)}{U_0}\right) dy , \qquad (5.34)$$

where y is the axis normal to the wall, $\langle U \rangle(y)$ [ms^{-1}] is the mean velocity at any height, and U_0 [ms^{-1}] is the free stream mean velocity. Alternatively, the *momentum thickness* can be defined as

$$\theta(x) \equiv \int_0^\infty \frac{\langle U \rangle(y)}{U_0} \left(1 - \frac{\langle U \rangle(y)}{U_0}\right) dy . \qquad (5.35)$$

Provide an argument to rank these thicknesses, i.e. $\delta(x)$, $\delta^*(x)$, and $\theta(x)$ [m] at any x [m], from largest to smallest. Likewise, various Reynolds numbers can be defined associated with each thickness quantity, such that

$$Re_\delta(x) = \frac{U_0 \delta(x)}{\nu}, \; Re_{\delta*}(x) = \frac{U_0 \delta^*(x)}{\nu}, \; Re_\theta(x) = \frac{U_0 \theta(x)}{\nu} . \qquad (5.36)$$

Provide an argument to rank these Reynolds numbers from largest to smallest.

5.5 A hypothetical boundary layer is characterized by the following velocity profile at position x [m].

$$\begin{cases} \langle U \rangle(y) = y \text{ ms}^{-1} & \text{for } y < 10 \, \text{m} , \\ \langle U \rangle(y) = U_0 = 10 \, \text{ms}^{-1} & \text{for } y \geq 10 \, \text{m} . \end{cases}$$

Using the definitions in the previous problem, calculate $\delta(x)$, $\delta^*(x)$, $\theta(x)$ [m], $Re_\delta(x)$, $Re_{\delta*}(x)$, and $Re_\theta(x)$ [−]. Can you confirm the ranks from largest to smallest hypothesized in the previous problem?

5.6 A large-eddy simulation (LES), to be introduced in Chap. 16, is used to model airflow inside a wind tunnel, as shown in Fig. 5.5. The tunnel has a smooth surface, and various vertical profiles of non-dimensional velocity U^+ [−] versus non-dimensional distance normal to the wall $(z - d)^+$ [−] are obtained at various distances downstream of the flow in the x direction. Here d [m] is the displacement height that is used to normalize normal distance to the wall. Usually d [m] is in the order of roughness characteristic length of a surface. For a smooth surface $d \approx 0$ m. Furthermore, the LES model is compared with experimental measurements of [6].

For each LES profile, provide approximate ranges of $(z - d)^+$ [−] associated with the inner layer and the outer layer. In addition, provide approximate ranges of $(z - d)^+$ [−] for the viscous sublayer, buffer sublayer, viscous wall region, overlap region, and the log-law region. Is there a unique slope that characterizes the log-law region? Using the experimental profile, explain what happens after the log-law region for very large values of $(z - d)^+$ [−]?

5.7 A mechanical engineer is analysing a hypothetical boundary layer characterized by the following velocity profile at position x [m]:

$$\begin{cases} \langle U \rangle(y) = y^{1/2} \text{ ms}^{-1} & \text{for } y < 9 \, \text{m} , \\ \langle U \rangle(y) = U_0 = 3 \, \text{ms}^{-1} & \text{for } y \geq 9 \, \text{m} , \end{cases}$$

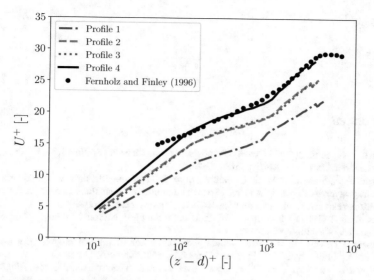

Fig. 5.5 Profiles of non-dimensional horizontal velocity versus non-dimensional distance normal to the wall for a smooth surface. Profiles 1, 2, 3, and 4 are obtained using a large-eddy simulation (LES) model that resolves flow near the walls. The experimental profile is obtained from [6]

where flow is in the x direction and y is the direction normal to the surface. Using the definition for the momentum thickness, $\theta(x)$ [m], and the Reynolds number associated with this thickness, $Re_\theta(x)$ [−], help this engineer calculate both $\theta(x)$ [m] and $Re_\theta(x)$ [−]. Assume that the kinematic viscosity of air is $\nu = 1.5 \times 10^{-5} \mathrm{m^2 s^{-1}}$.

5.8 Scales of turbulence near walls for most engineering applications are much finer than elsewhere in a system. Therefore, near walls, say in the log-law region, it can be assumed that the production rate and dissipation rate of turbulence kinetic energy are in equilibrium, i.e. they are the same. Suppose near a wall the mean flow is in the x direction with mean velocity $\langle U \rangle$ [ms^{-1}] and y is the direction normal to the wall. Also suppose there is horizontal homogeneity. The equivalency of production rate and dissipation rate of turbulence kinetic energy can be stated as

$$P_k = \epsilon , \tag{5.37}$$

where the production rate of turbulence kinetic energy can be expressed as $P_k = -\langle uv \rangle \frac{d\langle U \rangle}{dy}$ [m^{2}s^{-3}]. Here u and v [ms^{-1}] are turbulent fluctuations in the x and y directions. Assuming that the region of interest is the log-law region, express the Reynolds stress $\langle uv \rangle$ [m^{2}s^{-2}] as a function of friction velocity u_τ [ms^{-1}]. Also, assuming that the region of interest is the log-law region, express the gradient of mean velocity $\frac{d\langle U \rangle}{dy}$ [s^{-1}] as a function of friction velocity u_τ [ms^{-1}], von Kármán constant κ [−], and distance away from the wall y [m]. Using these assumptions, show that

$$\epsilon = \frac{u_\tau^3}{\kappa y} \, . \tag{5.38}$$

References

1. Kaimal J C, Finnigan J J (1994) Atmospheric boundary layer flows: their structure and measurement. Oxford University Press, Oxford
2. Aliabadi A A, Staebler R M, de Grandpré J et al (2016) Comparison of estimated atmospheric boundary layer mixing height in the Arctic and southern Great Plains under statically stable conditions: experimental and numerical aspects. Atmos Ocean 54:60–74
3. von Kármán T (1931) Mechanical similitude and turbulence. National Advisory Committee for Aeronautics, Washington DC
4. Bredberg J (2000) On the wall boundary condition for turbulence models. Department of Thermo and Fluid Dynamics, Chalmers University of Technology, Göteborg
5. Kays W M, Crawford M E (1993) Convective heat and mass transfer, 3rd edn. McGraw-Hill Inc., New York
6. Fernholz H H, Finley P J (1996) The incompressible zero-pressure-gradient turbulent boundary layer: An assessment of the data. Prog Aerospace Sci 32:245–311

Chapter 6
Free Shear Flows

Abstract This chapter introduces free shear flows, which are flows that are not bounded by any walls. The round jet is defined as a fundamental type of free shear flow. The axial, lateral, and radial components of velocity vector in a three-dimensional round jet are defined. The concept of self-similarity is introduced, which is used to empirically formulate expressions for velocity and Reynolds stress components of a round jet. The continuity and momentum transport equations for a two-dimensional jet are derived using simplifying assumptions.

6.1   Overview

The methodology developed in this chapter is adapted from [1]. In Chap. 5 we discussed flows that were bounded by a surface in at least one boundary. In this chapter we discuss free shear flows. Free shear flows are those that are remote from walls or surfaces. Typical examples include jets, wakes, and mixing layers. Turbulence in shear flows arises because of the existence of mean velocity gradients.

6.2   Round Jet

A common shear flow is the round jet, for instance encountered in internal combustion engines employing a nozzle to spray fuel into a combustion chamber [2] or a jet engine that propels an aircraft forward using a nozzle that ejects a combusted mixture of air and fuel [3]. This type of flow is created by a fluid exiting a nozzle with diameter d [m] that produces approximately a flat-topped and uniform velocity profile U_J [ms^{-1}]. The flow then enters an ambient background at rest. The flow is statistically stationary and axisymmetric. Therefore, the statistics of the flow would only depend on axial and radial coordinates, i.e. x and r [m] but are independent of time and circumferential coordinate, θ [Rad] [4]. The velocity components in the x, r, and θ coordinates are denoted by U, V, and W [ms^{-1}]. This is shown in Fig. 6.1.

© The Author(s), under exclusive license to Springer Nature Switzerland AG 2022
A. A. Aliabadi, *Turbulence*, Mechanical Engineering Series,
https://doi.org/10.1007/978-3-030-95411-6_6

Fig. 6.1 Schematic of a
round jet showing the
cylindrical coordinate system

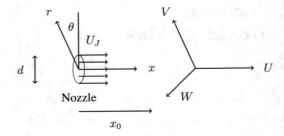

In the ideal scenario the jet can be specified by U_J [ms^{-1}], d [m], and ν [m^{2}s^{-1}]. The jet Reynolds number can also be defined using these parameters as

$$Re = \frac{U_J d}{\nu} . \tag{6.1}$$

However, in practice the nozzle dimension and the surroundings have some effect on the jet, so more specifications can be given to characterize a jet [5].

The mean velocity of a jet is predominantly in the axial direction. Note that at a radial distance of $r = 0$ m the jet axis is defined, about which the profiles of the mean axial velocity $\langle U \rangle$ [ms^{-1}] are symmetric. Due to symmetry, the mean circumferential velocity is zero, i.e. $\langle W \rangle = 0$ ms^{-1}, while the mean radial velocity is not zero but an order of magnitude smaller than the mean axial velocity, i.e. $\langle V \rangle \ll \langle U \rangle$ [ms^{-1}].

6.3   Axial Velocity

The centre-line mean axial velocity can be defined in terms of the mean axial velocity, $\langle U(x, r, \theta) \rangle$ [ms^{-1}] such that

$$U_0(x) \equiv \langle U(x, 0, 0) \rangle . \tag{6.2}$$

The jet's half-width $r_{1/2}(x)$ [m] is defined as a radial distance where the mean axial velocity is half of the centre-line mean axial velocity, i.e.

$$\langle U(x, r_{1/2}(x), 0) \rangle \equiv \frac{1}{2} U_0(x) . \tag{6.3}$$

Note that the jet's half-width is not constant but a function of axial distance. In fact this half-width constant increases further away from the nozzle.

6.4 Self-similarity

Self-similarity is an important concept in the study of turbulent flows. In mathematics, a self-similar object is exactly or approximately similar to a part of itself (i.e. the whole has the same shape as one or more of the parts). Many objects in the real world, such as coastlines and trees, are statistically self-similar: parts of them show the same statistical properties at many scales. Figure 6.2 shows a self-similar object that resembles a tree.

Consider a quantity $Q(x, y)$ [–] that depends on two independent variables x and y [–]. As functions of x [–], characteristic scales $Q_0(x)$ [–] and $\delta(x)$ [–] are defined for the dependent variable Q [–] and the independent variable y [–], respectively. Then scaled variables are defined by

$$\xi \equiv \frac{y}{\delta(x)} , \tag{6.4}$$

$$\tilde{Q}(\xi, x) \equiv \frac{Q(x, y)}{Q_0(x)} . \tag{6.5}$$

If the scaled dependent variable is independent of x [–], i.e. there is a function $\hat{Q}(\xi)$ [–] such that

$$\tilde{Q}(\xi, x) = \hat{Q}(\xi) , \tag{6.6}$$

then $Q(x, y)$ [–] is self-similar. In this case, $Q(x, y)$ [–] can be expressed in terms of functions of single independent variables, $Q_0(x)$ [–], $\delta(x)$ [–], and $\hat{Q}(\xi)$ [–]. In a successful formulation of self-similar phenomena, it is important to choose scales $Q_0(x)$ [–] and $\delta(x)$ [–] appropriately. In some circumstances, more general transformations are required. It is also important to note that self-similar behaviour

Fig. 6.2 A self-similar object

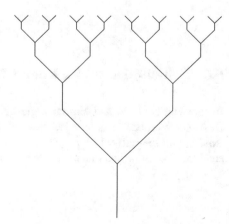

may not be observed over the entire range of independent variables, but only for a limited range.

6.5   Axial Variation of Scales

To characterize a jet we need to determine variations of two scales, i.e. the velocity scale $U_0(x)$ [ms^{-1}] and lengthscale $r_{1/2}(x)$ [m]. It is experimentally verified that the inverse of $U_0(x)$ [ms^{-1}], specifically $U_J/U_0(x)$ [–], plotted against x/d [–], falls on a straight line. The intercept of this line with the abscissa defines the *virtual origin*, denoted by x_0 [m], so that the straight line corresponds to

$$\frac{U_0(x)}{U_J} = \frac{B}{(x - x_0)/d} ,$$
(6.7)

where $B \approx 5.9$ [–] is an empirical constant. It has been experimentally observed that $x_0/d \approx 4$ [–] [5]. The virtual origin x_0 [m] represents, mathematically, the location where the self-similar behaviour of a jet begins.

The *spreading* of a jet characterizes how the jet grows in the radial direction as a function of axial distance away from the nozzle. The jet *spreading rate* can be defined using the differential equation

$$S \equiv \frac{dr_{1/2}(x)}{dx} ,$$
(6.8)

which is a constant quantity in the self-similar region of the jet. The spreading of a jet has been measured experimentally as $S \approx 0.102$ [–] [5]. Integrating the spreading rate (Eq. 6.8), we can obtain a formula for the jet's half-width in the self-similar region

$$r_{1/2}(x) = S(x - x_0) .$$
(6.9)

6.6   Self-similarity of a Round Jet

In the well-behaved self-similar region of a round jet ($x/d > 30$ [–]) and a high Reynolds number turbulent jet ($Re > 10^4$ [–]), the centre-line velocity $U_0(x)$ [ms^{-1}] and the half-width $r_{1/2}(x)$ [m] vary according to equations in the previous section. The empirical constants, B [–] and S [–], are independent of the Reynolds number. A cross-stream *similarity variable* can be taken as either of

$$\xi \equiv \frac{r}{r_{1/2}} \, , \tag{6.10}$$

$$\eta \equiv \frac{r}{x - x_0} \, . \tag{6.11}$$

These two variables are in fact related by $\eta = S\xi$ [–]. The self-similar mean velocity profile can be defined by

$$f(\eta) = \overline{f}(\xi) = \frac{\langle U(x, r, 0)\rangle}{U_0(x)} \, , \tag{6.12}$$

where functions can be fitted for η [–]. An approximation is $f(\eta) = (1 + a\eta^2)^{-2}$ [–], where a [–] is a constant. The mean lateral velocity in the self-similar region, i.e. $\langle V \rangle$ [ms^{-1}], can be determined from $\langle U \rangle$ [ms^{-1}] via the continuity Eq. 2.2. Likewise, self-similar profiles of $\langle V \rangle / U_0$ [–] can be obtained. The mean lateral velocity is usually found to be more than one order of magnitude smaller than the mean axial velocity, i.e. $|\langle V \rangle| \approx 0.03|\langle U \rangle|$ [ms^{-1}]. At the edge of the jet, the mean lateral velocity is negative, i.e. the ambient fluid is being pulled towards the centre-line axis. This is known as *entrainment*.

6.7 Reynolds Stresses

For a round jet, the fluctuating velocity components in the x, r [m], and θ [Rad] coordinate directions are denoted by u, v, and w [ms^{-1}]. The Reynolds stress tensor is

$$\begin{bmatrix} \langle u^2 \rangle & \langle uv \rangle & 0 \\ \langle uv \rangle & \langle v^2 \rangle & 0 \\ 0 & 0 & \langle w^2 \rangle \end{bmatrix} ,$$

where due to circumferential symmetry, $\langle uw \rangle$ and $\langle vw \rangle$ [m^{2}s^{-2}] are zero. The geometry of the flow dictates that the normal stresses be even functions of r [m], while the shear stress $\langle uv \rangle$ [m^{2}s^{-2}] is an odd function. As the r [m] approaches zero, the radial V [ms^{-1}] and circumferential W [ms^{-1}] components of velocity become indistinguishable. Therefore on the axis of the jet $\langle v^2 \rangle$ and $\langle w^2 \rangle$ [m^{2}s^{-2}] become equal.

Consider the root mean square of axial velocity on the centre-line defined as

$$u'_0(x) \equiv \langle u^2 \rangle_{r=0}^{1/2} \, . \tag{6.13}$$

Table 6.1 Constants to determine turbulent properties of a self-similar round jet [5]

$p(\eta)$	C_0	C_2	C_4	C_6	A
$\langle U \rangle / U_0(x)$	1.0	-1.925	0.0	0.0	63
$\langle u^2 \rangle / U_0^2(x)$	7.778e$-$2	2.79e1	-2.02e3	4.3e5	257
$\langle v^2 \rangle / U_0^2(x)$	5.457e$-$2	0.355	-4.298e1	0.0	89
$\langle w^2 \rangle / U_0^2(x)$	5.78e$-$2	-1.71	2.73e$-$1	0.0	42
$\langle uv \rangle / U_0^2(x)$	4.375e$-$1	-3.931e1	1.55e2	1.342e4	90
$\epsilon / \left[U_0^3(x)/(x - x_0) \right]$	0.3549	11.99	-1635	43470	201

We wish to characterize how $u_0'(x)$ [ms^{-1}] will vary as a function of x [m], or equivalently in terms of non-dimensional quantities, how the ratio $u_0'(x)/U_0(x)$ [–] varies with x/d [–]. It has been observed that after the development region of the jet, i.e. after the virtual origin, both $u_0'(x)$ and $U_0(x)$ [ms^{-1}] decay as x^{-1} [m^{-1}].

In a similar way to mean velocities, it has been found that the Reynolds stresses become self-similar. That is, profiles of $\langle u_i u_j \rangle / U_0^2(x)$ [–] plotted against $r/r_{1/2}$ [–] or $\eta \equiv r/(x - x_0)$ [–] collapse for all x [m] beyond the development region on the same curve.

The Reynolds stress exhibits significant anisotropy. The relative magnitude of the shear stress compared to the normal stresses can be quantified using $\langle uv \rangle / k \approx 0.27$ [–]. The velocity fluctuations of u and v [ms^{-1}] are found to be correlated with a correlation coefficient of $\rho_{uv} \approx 0.4$ [–].

Numerous functions have been fitted to describe round jets, such as those provided by [5], where model fitting to experimental data is achieved by the method of least squares to all measured profiles with a similarity variable and fitted an even function given below to calculate mean flow and turbulent properties,

$$p(\eta) = \left[C_0 + C_2 \eta^2 + C_4 \eta^4 + \ldots \right] \exp \left(-A \eta^2 \right) . \tag{6.14}$$

The multiplication of the polynomial and exponential functions provides an excellent fit over the range in which data were taken ($\eta \equiv r/(x - x_0) < 0.2$ [–]), and care must be given not to apply these fits beyond this range. Table 6.1 shows the fitting parameters.

6.8 Mean Continuity and Momentum Equations for a Jet

We wish to derive the mean continuity and momentum equations for a two-dimensional jet. The jet axis is in the x direction and the lateral direction is y. Far away from the jet in the lateral direction $y \rightarrow \infty$ [m], the fluid is quiescent at a pressure p_0 [kgm^{-1}s^{-2}] [6]. The continuity equation is given as

$$\frac{\partial \langle U \rangle}{\partial x} + \frac{\partial \langle V \rangle}{\partial y} = 0 . \tag{6.15}$$

The momentum equations in the x and y directions can be written as follows [1]. Note that terms in curly brackets {} can be assumed negligible.

$$\underbrace{\langle U \rangle \frac{\partial \langle U \rangle}{\partial x} + \langle V \rangle \frac{\partial \langle U \rangle}{\partial y}}_{\text{Advection}} = \underbrace{\{\nu \frac{\partial^2 \langle U \rangle}{\partial x^2}\} + \nu \frac{\partial^2 \langle U \rangle}{\partial y^2}}_{\text{Surface Force}} - \underbrace{\frac{\partial \langle uv \rangle}{\partial y}}_{\text{Shear Stress Force}}$$

$$- \underbrace{\{\frac{\partial \langle u^2 \rangle}{\partial x}\}}_{\text{Normal Stress Force}} - \underbrace{\frac{1}{\rho} \frac{\partial \langle p \rangle}{\partial x}}_{\text{Normal and Body Forces}},$$

$$\tag{6.16}$$

$$\underbrace{\{\langle U \rangle \frac{\partial \langle V \rangle}{\partial x}\} + \{\langle V \rangle \frac{\partial \langle V \rangle}{\partial y}\}}_{\text{Advection}} = \underbrace{\{\nu \frac{\partial^2 \langle V \rangle}{\partial x^2}\} + \{\nu \frac{\partial^2 \langle V \rangle}{\partial y^2}\}}_{\text{Surface Force}} - \underbrace{\{\frac{\partial \langle uv \rangle}{\partial x}\}}_{\text{Shear Stress Force}}$$

$$- \underbrace{\frac{\partial \langle v^2 \rangle}{\partial y}}_{\text{Normal Stress Force}} - \underbrace{\frac{1}{\rho} \frac{\partial \langle p \rangle}{\partial y}}_{\text{Normal and Body Forces}}.$$

$$\tag{6.17}$$

The rationale for the neglected terms {} can be explained conveniently. All terms containing a partial derivative of $\langle V \rangle$ [ms^{-1}] are neglected since mean velocity in the lateral direction is small. Also all partial derivatives for components of Reynolds stress, involving both normal and shear stresses, with respect to x are neglected because these derivatives are insignificant with respect to derivatives in the y direction. Finally, second derivatives with respect to x are neglected since variations in the axial direction exhibit negligible curvature. The mean lateral momentum Eq. 6.17 simplifies to

$$\frac{1}{\rho} \frac{\partial \langle p \rangle}{\partial y} + \frac{\partial \langle v^2 \rangle}{\partial y} = 0 , \tag{6.18}$$

which can be integrated from an arbitrary y to $y \to \infty$ [m] to give an expression relating mean pressure to far field pressure p_0 [kgm^{-1}s^{-2}] and $\langle v^2 \rangle$ [m^{2}s^{-2}] such that

$$\langle p \rangle / \rho = p_0 / \rho - \langle v^2 \rangle . \tag{6.19}$$

This equation can now be differentiated with respect to x to give the axial pressure gradient as a function of far field pressure p_0 [kgm^{-1}s^{-2}] and $\langle v^2 \rangle$ [m^{2}s^{-2}]

such that

$$\frac{1}{\rho}\frac{\partial \langle p \rangle}{\partial x} = \{\frac{1}{\rho}\frac{dp_0}{dx}\} - \{\frac{\partial \langle v^2 \rangle}{\partial x}\} \approx 0 , \tag{6.20}$$

which is valid since in a quiescent background p_0 [kgm^{-1}s^{-2}] is uniform and that we already assumed partial derivatives of Reynolds stress components in the x direction are negligible. These developments result in a much shorter version of the mean momentum Eq. 6.16 in the axial direction given by

$$\underbrace{\langle U \rangle \frac{\partial \langle U \rangle}{\partial x} + \langle V \rangle \frac{\partial \langle U \rangle}{\partial y}}_{\text{Advection}} = \underbrace{v \frac{\partial^2 \langle U \rangle}{\partial y^2}}_{\text{Surface Force}} - \underbrace{\frac{\partial \langle uv \rangle}{\partial y}}_{\text{Shear Stress Force}} . \tag{6.21}$$

It is important to note that this equation does not have a pressure term, suggesting that the pressure variation in a jet with quiescent background is small so that all transport of momentum is performed by advection, surface, and shear stress forces.

Problems

6.1 A round steady jet is produced by a nozzle with diameter $d = 0.01$ m. The jet exit velocity is $U_J = 1$ms^{-1}. Making the necessary assumptions, at position $x = 0.5$ m, show that the jet's mean axial velocity on the centre-line $U_0(x)$ [ms^{-1}] and the jet's half-width $r_{1/2}(x)$ [m] are given by

$$U_0(x = 0.5\,\text{m}) = 0.13\,\text{ms}^{-1} , \tag{6.22}$$

$$r_{1/2}(x = 0.5\,\text{m}) = 0.047\,\text{m} . \tag{6.23}$$

6.2 In the previous problem, show that the mean velocity in the axial direction at $x = 0.5$ m and $r = 0.02$ m is given by

$$\langle U \rangle (x = 0.5\,\text{m}, r = 0.02\,\text{m}) = 0.12\,\text{ms}^{-1} . \tag{6.24}$$

6.3 The Fibonacci sequence is the series of numbers $0, 1, 1, 2, 3, 5, 8, 13, 21, 34, \ldots$ where the next number in the series is found by adding up the two numbers before it, i.e.

$$x_n = x_{n-1} + x_{n-2} . \tag{6.25}$$

This series exhibits self-similarity because each term depends on part of previous terms. A surprising result is that we can calculate any term of this series using the Golden Ratio $\phi = 1.618034\ldots$ such that

$$x_n = \frac{\phi^n - (1 - \phi)^n}{\sqrt{5}} . \tag{6.26}$$

For example,

$$x_6 = \frac{1.618034\ldots^6 - (1 - 1.618034\ldots)^6}{\sqrt{5}} = 8 \ .$$ (6.27)

So it appears that the Golden Ratio is a number in nature that describes self-similarity! In mathematics, two quantities form a golden ratio if their ratio is the same as the ratio of their sum to the larger of the two quantities. Expressed algebraically, for quantities a and b with $a > b > 0$, they form a Golden Ratio if

$$\frac{a+b}{a} = \frac{a}{b} \equiv \phi \ .$$ (6.28)

Use basic algebra to show that

$$\phi = \frac{1 + \sqrt{5}}{2} \ .$$ (6.29)

References

1. Pope S B (2000) Turbulent flows. Cambridge University Press, Cambridge
2. Heywood J B (1988) Internal combustion engine fundamentals. McGraw-Hill Inc, New York
3. White F M (2003) Fluid mechanics. McGraw-Hill Higher Education, New York
4. Aliabadi A A, Lim K W, Rogak S N et al (2011) Steady and transient droplet dispersion in an air-assist internally mixing cone atomizer. Atomization Spray 21:1009–1031
5. Hussein H J, Capp S P, George W K (1994) Velocity measurements in a high-Reynolds-number momentum-conserving, axisymmetric, turbulent jet. J Fluid Mech 258:31–75
6. Acheson D J (1990) Elementary fluid dynamics. Oxford University Press, Oxford

Chapter 7
Compressible Flows

Abstract This chapter introduces compressible flows, which are flows at very high velocities comparable to the speed of sound or flows influenced by temperature-driven buoyant forces. Transport equations are developed for continuity, momentum, and temperature, which link various non-constant flow properties such as density, velocity components, pressure, and temperature. The similitude analysis is performed, and non-dimensional transport equations are derived. Boundary layer equations are developed for a two-dimensional case of a compressible turbulent flow. Finally, the Mach number is defined, which characterizes the flow regimes in relation to the speed of sound, concerning subsonic, transonic, supersonic, hypersonic, and re-entry flows.

7.1 Overview

In the previous chapters we treated density as a constant fluid property in many instances. Further, we treated temperature as a scalar field, meaning that it cannot affect the flow velocity field. These assumptions are far from reality. Many flows subject to body forces exhibit density variations that are sensitive to small temperature changes. Such flows are buoyancy driven, with the buoyant force not negligible in impacting the flow velocity field [1]. Other flows may exhibit high velocities, with the consequence that the flow velocity, density, temperature, viscosity, pressure, and other properties would be related to one another [2]. In all such cases more complete system of transport equations should be used that link various flow properties appropriately. We treat this situation using the principles governing compressible flows. The methodology developed in this chapter is adapted from [3].

© The Author(s), under exclusive license to Springer Nature Switzerland AG 2022
A. A. Aliabadi, *Turbulence*, Mechanical Engineering Series,
https://doi.org/10.1007/978-3-030-95411-6_7

7.2 Continuity Equation for Compressible Flows

Chapter 2 Eq. 2.1 provided the general condition for continuity. This equation can also be used as the continuity criteria for compressible flows, in which density is not constant

$$\underbrace{\frac{\partial \rho}{\partial t}}_{\text{Storage}} + \underbrace{\nabla.(\rho \mathbf{U})}_{\text{Advection}} = 0 . \tag{7.1}$$

7.3 Energy Equation for Compressible Flows

We begin the theory by developing the energy equation for a fluid element in motion. Suppose that the volume and mass of the fluid element in motion, shown in Fig. 7.1, are given by $dV = dxdydz$ [m^3] and $dM = \rho dV$ [kg], respectively. Since the fluid element is moving, the first law of thermodynamics should be expressed using the material derivative for an enclosed parcel of moving fluid. The first law of thermodynamics can be written as

$$\underbrace{\frac{DQ}{Dt}}_{\text{Heat}} = \underbrace{\frac{DE_T}{Dt}}_{\text{Energy}} + \underbrace{\frac{DW}{Dt}}_{\text{Work}} , \tag{7.2}$$

Fig. 7.1 Control volume of a fluid element with demonstration of selected surface forces expressed as stress tensor components in a Cartesian coordinate system

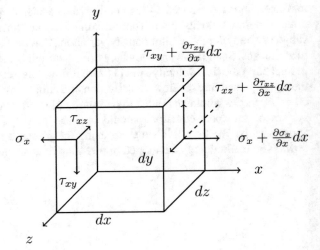

where the energy balance states that the heat Q [J] transferred into the fluid parcel must be equal to the change in its total energy E_T [J] plus the work W [J] of expansion (or compression).

If we neglect radiation heat transfer, then the transportation of heat can only occur via storage, convection, and conduction. The Fourier's law of heat conduction can be used to characterize the conduction of heat in any given direction through the moving fluid control volume dV [m^3]. For instance the amount of heat transferred into the control volume in the x direction is equal to $-k\frac{\partial T}{\partial x}dydz$ [W], where k [Wm^{-1}K^{-1}] is thermal conductivity. By contrast, the amount of heat leaving the control volume in the x direction is equal to $[k\frac{\partial T}{\partial x} + \frac{\partial}{\partial x}(k\frac{\partial T}{\partial x})dx]dydz$ [W]. Therefore, the net heat conduction in the x direction is $dV\frac{\partial}{\partial x}(k\frac{\partial T}{\partial x})$ [W]. So the overall heat balance for the moving fluid element can be written as

$$\frac{DQ}{Dt} = dV\nabla.(k\nabla T) . \tag{7.3}$$

The change in the total energy of the moving fluid parcel consists of the change in its internal energy $dE = \rho dV de$ [J] and the change in the kinetic energy $d[\frac{1}{2}\rho dV(U^2 + V^2 + W^2)]$ [J]. If we ignore the change in potential energy due to a displacement in the gravitational field, we can write

$$\frac{DE_T}{Dt} = \rho dV \left[\frac{De}{Dt} + \frac{1}{2}\frac{D}{Dt}\left(U^2 + V^2 + W^2\right) \right] . \tag{7.4}$$

To express the change in the work of expansion (or compression), we shall first review the tensor properties introduced in Chap. 4. The normal and shear stresses in the flow can be written as

$$\begin{bmatrix} \sigma_x & \tau_{xy} & \tau_{xz} \\ \tau_{yx} & \sigma_y & \tau_{yz} \\ \tau_{zx} & \tau_{zy} & \sigma_z \end{bmatrix} = -\rho \begin{bmatrix} \langle u^2 \rangle & \langle uv \rangle & \langle uw \rangle \\ \langle vu \rangle & \langle v^2 \rangle & \langle vw \rangle \\ \langle wu \rangle & \langle wv \rangle & \langle w^2 \rangle \end{bmatrix} . \tag{7.5}$$

In the definition of first law of thermodynamic provided, the work added to the fluid from the outside is negative. To determine the work performed, we must calculate the contribution of all stress components on each surface of the fluid control volume. Consider the contribution of the component σ_x [kgm^{-1}s^{-2}] of stress. Considering a pair of surfaces normal to the x axis, the work exerted per unit time can be calculated by accounting for the displacement of each surface via the corresponding velocities and the normal stress

$$\frac{DW_{\sigma_x}}{Dt} = -dydz\left[-U\sigma_x + \left(U + \frac{\partial U}{\partial x}dx\right)\left(\sigma_x + \frac{\partial \sigma_x}{\partial x}dx\right) \right]$$

$$= -dV\frac{\partial}{\partial x}(U\sigma_x) . \tag{7.6}$$

Using this logic, the total work on the fluid element can be calculated considering all the normal and shear stresses on each pair of fluid element surfaces. The work can be written as

$$
\frac{DW}{Dt} = -dV \left\{ \frac{\partial}{\partial x} (U\sigma_x + V\tau_{xy} + W\tau_{xz}) \right.
$$

$$
+ \frac{\partial}{\partial y} (U\tau_{yx} + V\sigma_y + W\tau_{yz})
$$

$$
\left. + \frac{\partial}{\partial z} (U\tau_{zx} + V\tau_{zy} + W\sigma_z) \right\} . \tag{7.7}
$$

Substituting Eqs. 7.3, 7.4, 7.7 into Eq. 7.2 and performing a number of simplifications, we can obtain the energy equation for compressible flows, while neglecting radiation heat transfer and potential energy, such that

$$
\rho \frac{De}{Dt} + P\nabla.\mathbf{U} = \nabla.(k\nabla T) + \mu\Phi , \tag{7.8}
$$

where P [Pa] is the pressure, $\mathbf{U}$ [ms^{-1}] is the flow velocity vector, μ [kgm^{-1}s^{-1}] is the dynamic viscosity, and Φ [s^{-2}] represents the dissipation function given by

$$
\Phi = 2 \left[\left(\frac{\partial U}{\partial x}\right)^2 + \left(\frac{\partial V}{\partial y}\right)^2 + \left(\frac{\partial W}{\partial z}\right)^2 \right]
$$

$$
+ \left(\frac{\partial V}{\partial x} + \frac{\partial U}{\partial y}\right)^2 + \left(\frac{\partial W}{\partial y} + \frac{\partial V}{\partial z}\right)^2 + \left(\frac{\partial U}{\partial z} + \frac{\partial W}{\partial x}\right)^2
$$

$$
- \frac{2}{3} \left(\frac{\partial U}{\partial x} + \frac{\partial V}{\partial y} + \frac{\partial W}{\partial z}\right)^2 . \tag{7.9}
$$

Equation 7.8 can be simplified for the case of an ideal gas, which represents a compressible fluid. We first consider that the material derivative of density and the divergence of velocity vector are related. Starting from the continuity Eq. 2.1, we can show that

$$
\frac{\partial \rho}{\partial t} + \nabla.(\rho \mathbf{U}) = 0
$$

$$
\underbrace{\frac{\partial \rho}{\partial t} + \mathbf{U}.\nabla \rho}_{\frac{D\rho}{Dt}} + \rho \nabla.\mathbf{U} = 0 . \tag{7.10}
$$

We can next consider basic thermodynamic relationships for the ideal gas. Over a finite range of temperatures, we can assume that the heat capacity of an ideal gas under constant volume c_v [Jkg^{-1}K^{-1}] is constant, such that $de = c_v dT$ [Jkg^{-1}]. Further, we can assume that the heat capacity of an ideal gas under constant pressure c_p [Jkg^{-1}K^{-1}] is also constant, such that $dh = c_p dT$ [Jkg^{-1}], where $h = e + \frac{P}{\rho}$ [Jkg^{-1}] is enthalpy. In fact the differentials of enthalpy and internal energy are related via

$$dh = de + d\left(\frac{P}{\rho}\right)$$

$$c_p dT = c_v dT + \frac{1}{\rho}dP - \frac{P}{\rho^2}d\rho . \tag{7.11}$$

Now Eqs. 7.10 and 7.11 can be used to simplify Eq. 7.8 to a form that is general for ideal gases, which can describe the compressible behaviour of many fluids

$$\underbrace{\rho c_p \frac{DT}{Dt}}_{\text{Material Derivative}} = \underbrace{\frac{DP}{Dt}}_{\text{Pressure}} + \underbrace{\nabla.(k\nabla T)}_{\text{Conduction}} + \underbrace{\mu\Phi}_{\text{Dissipation}} . \tag{7.12}$$

The temperature variations by pressure variations in compressible flow are important aspects of design. This can be demonstrated by an idealized case, where a flow parcel is adiabatically and reversibly compressed along a fluid streamline. That means that the conduction and viscous dissipation terms can be neglected for this purpose. Also to simplify our calculation we can consider a one-dimensional case, where the streamline and fluid velocity are only along the x axis. In this case the only variable of interest becomes the stream variable s [–] that identifies the position of interest along the streamline. Our goal is to find the change in flow properties along the streamline from U_∞ [ms^{-1}], T_∞ [K], and P_∞ [Pa] to U_0 [ms^{-1}], T_0 [K], and P_0 [Pa]. In this simplification only the advective terms remain in the material derivatives in Eq. 7.12

$$\rho c_p U \frac{dT}{ds} = U \frac{dP}{ds} . \tag{7.13}$$

Simplifying and integrating this equation from $s = \infty$ to $s = 0$ [–] and using the Bernoulli equation (permitted for the idealized case here), we can arrive at

$$T_0 - T_\infty = \frac{1}{c_p}\int_{P_\infty}^{P_0} \frac{dP}{\rho} = \frac{U_\infty^2 - U_0^2}{2c_p} . \tag{7.14}$$

This is an important design equation in aerodynamics. If the end point along the streamline brings the fluid (air) to a complete stop ($U_0 = 0$ ms^{-1}), the resulting temperature rise is known as the stagnation temperature rise. This temperature rise gives an estimate of increase in temperature around various segments of a moving object that brings the air to a complete stop with respect to the object. For a moving automobile at $U_\infty = 40$ ms^{-1} we obtain $T_0 - T_\infty = 1$ K. For a fast moving supersonic aircraft at $U_\infty = 500$ ms^{-1} we obtain $T_0 - T_\infty = 100$ K.

7.4   Momentum Equation for Compressible Flows

In the previous section we provided the energy equation for compressible flows characterized as ideal gases. In this section we will formulate the remaining transport equation of momentum for compressible flows characterized as ideal gases. In flows where temperature differences cause density differences, it is necessary to include buoyancy forces in the equations of motion along directions where body forces (e.g. gravitational and centrifugal forces) are present. In fact the buoyancy force itself is a body force that arises due to density differences accompanied by a pre-existing body force. The magnitude of the buoyancy term in the momentum equations along each of the coordinate axes x, y, and z is proportional to ρg_x, ρg_y, and ρg_z [kgm^{-2}s^{-2}], where g_x, g_y, and g_z [ms^{-2}] are the body accelerations due to body forces along each direction.

For ideal gases, the density is a function of both pressure and temperature via the ideal gas law $P = \rho R T$ [Pa], where P [Pa] is the pressure, ρ [kgm^{-3}] is the density, R [Jkg^{-1}K^{-1}] is the specific gas constant, and T [K] is the temperature . Using this law and the Taylor series expansion around reference density ρ_∞ [kgm^{-3}], pressure P_∞ [Pa], and temperature T_∞ [K], we can write

$$
\begin{aligned}
\rho &= \rho_\infty + \left(\frac{\partial \rho}{\partial T}\right)_{P_\infty, T_\infty} (T - T_\infty) + \left(\frac{\partial \rho}{\partial P}\right)_{P_\infty, T_\infty} (P - P_\infty) \\
&= \rho_\infty - \frac{P_\infty}{RT_\infty^2}(T - T_\infty) + \frac{1}{RT_\infty}(P - P_\infty) \\
&= \rho_\infty - \rho_\infty \beta (T - T_\infty) + \frac{\gamma}{c_\infty^2}(P - P_\infty) ,
\end{aligned}
\tag{7.15}
$$

where $\beta = \frac{1}{T_\infty}$ [K^{-1}] is the coefficient of thermal expansion, and $c_\infty = \sqrt{\gamma R T_\infty}$ [ms^{-1}] is the speed of sound for an ideal gas at temperature T_∞ [K]. $\gamma = \frac{c_p}{c_v}$ [–] is the ratio of specific heats. The last term shall be neglected in flows that are affected by body forces. This means that the dependency of density on pressure can be ignored in comparison to the dependency of density on temperature.

It can be remembered from Chap. 2 Eq. 2.7 that the modified pressure can be defined in such a way to absorb the field potential of a body force (e.g. gravitational or centrifugal body forces). Alternatively, the term associated with the body force may not be absorbed in the pressure and instead may be written as a standalone term. Using this convention, the momentum equation for a compressible flow of constant viscosity can be written as

$$\underbrace{\rho \frac{D\mathbf{U}}{Dt}}_{\text{Material Derivative}} = - \underbrace{\nabla P}_{\text{Surface Normal Force}} + \underbrace{\rho \mathbf{g} - \rho\beta(T - T_\infty)\mathbf{g}}_{\text{Body Forces}} \quad (7.16)$$

$$+ \underbrace{\mu\nabla^2\mathbf{U}}_{\text{Surface Shear Force}} ,$$

where P [Pa] is the flow pressure and $\mathbf{g}$ [ms^{-2}] is the overall body acceleration. It can be seen that the body force $\rho\mathbf{g}$ [kgm^{-2}s^{-2}] is aligned with the body acceleration vector, but the buoyant force $-\rho\beta(T - T_\infty)\mathbf{g}$ [kgm^{-2}s^{-2}] is in the opposite direction of the body acceleration. For instance in atmospheric flows the gravitational acceleration points down towards the earth surface (negative), while a warm parcel of air $T > T_\infty$ [K] experiences an upward buoyant force (positive).

As discussed in the above sections, the compressible flows require six simultaneous equations to solve for six variables: U, V, W [ms^{-1}], P [Pa], ρ [kgm^{-3}], and T [K]. The corresponding equations available are the continuity Eq. 2.1 (1), momentum Eq. 7.17 (3), the energy Eq. 7.12 (1), and the ideal gas law (1).

7.5 Similitude and Non-dimensional Transport Equations for Compressible Flows

To pursue the same tradition as in Chap. 2, we wish to use the concept of *similitude* to find the non-dimensional groups that govern the behaviour of the transport equations in the compressible flow [4]. Consider a system with characteristic length $\mathcal{L}$ [m] and characteristic velocity $\mathcal{U}_\infty$ [ms^{-1}], associated with an object immersed in flow. Further assume ρ_∞ [kgm^{-3}] and T_∞ [K] to be associated with density and temperature of the fluid, respectively, far away from the object. We can consider T_w [K] to represent the temperature on the skin of the object. Then, numerous non-dimensional *independent* and *dependent variables* can be formed to be used in the transport equations

$$\hat{\mathbf{x}} = \frac{\mathbf{x}}{\mathcal{L}} , \tag{7.17}$$

$$\hat{t} = \frac{t\mathcal{U}_\infty}{\mathcal{L}} , \tag{7.18}$$

$$\hat{\mathbf{U}}(\hat{\mathbf{x}}, \hat{t}) = \frac{\mathbf{U}(\mathbf{x}, t)}{\mathcal{U}_\infty} , \tag{7.19}$$

$$\hat{p}(\hat{\mathbf{x}}, \hat{t}) = \frac{p(\mathbf{x}, t)}{\rho_\infty \mathcal{U}_\infty^2} \, , \tag{7.20}$$

$$\hat{\rho} = \frac{\rho}{\rho_\infty} \, , \tag{7.21}$$

$$\hat{T} = \frac{T - T_\infty}{T_w - T_\infty} = \frac{T - T_\infty}{(\Delta T)_0} \, . \tag{7.22}$$

To continue the similitude analysis we revert to using the modified pressure p [kgm^{-1}s^{-2}] as opposed to the total pressure P [kgm^{-1}s^{-2}]. For the momentum equation, this is equivalent to subtracting the term $\nabla P_{st} = \rho_\infty \mathbf{g}$ [kgm^{-2}s^{-2}], generated by the static field, from the right hand side of the momentum equation.

It must be noted that with variable transformations, the total derivative, divergence, gradient, Laplacian, and other operators will all be defined over the transformed independent and dependent variables, i.e. $\hat{\mathbf{x}}$ [–], $\hat{t}$ [–], and $\hat{\mathbf{U}}$ [–]. For instance $\nabla^2 \hat{T} = \frac{\partial \hat{T}}{\partial \hat{x}^2} + \frac{\partial \hat{T}}{\partial \hat{y}^2} + \frac{\partial \hat{T}}{\partial \hat{z}^2}$ and $\frac{D\hat{T}}{D\hat{t}} = \frac{\partial \hat{T}}{\partial \hat{t}} + \hat{U}\frac{\partial \hat{T}}{\partial \hat{x}} + \hat{V}\frac{\partial \hat{T}}{\partial \hat{y}} + \hat{W}\frac{\partial \hat{T}}{\partial \hat{z}}$. The continuity equation in the non-dimensional form using similitude can be expressed as

$$\frac{\partial \hat{\rho}}{\partial \hat{t}} + \nabla.(\hat{\rho}\hat{\mathbf{U}}) = 0 \, , \tag{7.23}$$

which is a familiar result, similar to the incompressible case. For the energy equation we can replace the total pressure with the modified pressure and assume thermal conductivity to be constant such that

$$\hat{\rho}\frac{D\hat{T}}{D\hat{t}} = \underbrace{\frac{U_\infty^2}{c_p(\Delta T)_0}}_{Ec} \frac{D\hat{p}}{D\hat{t}} + \underbrace{\frac{k}{\rho_\infty c_p U_\infty \mathcal{L}}}_{\frac{1}{Re \times Pr}} \nabla^2 \hat{T} + \underbrace{\frac{\mu U_\infty}{\rho_\infty c_p \mathcal{L}(\Delta T)_0}}_{\frac{Ec}{Re}} \hat{\Phi} \, , \tag{7.24}$$

where $\hat{\Phi} = 2\left[\left(\frac{\partial \hat{U}}{\partial \hat{x}}\right)^2 + \ldots\right] + \ldots$. $Re = \frac{\rho_\infty U_\infty \mathcal{L}}{\mu}$ is the familiar Reynolds number.

$Ec = \frac{U_\infty^2}{c_p(\Delta T)_0}$ [–] is known as the Eckert number, which expresses the relationship between a flow's kinetic energy and enthalpy, and is used to characterize dissipation. $Pr = \frac{\mu c_p}{k}$ [–] is the familiar Prandtl number, which characterizes the relationship between viscous and thermal diffusion rates. The product $Pe = Re \times Pr$ [–] is also known as the Péclet number, which signifies the relative importance of convection versus conduction mechanisms [5].

For the momentum equation, again, we can replace the total pressure with the modified pressure and assume dynamic viscosity to be constant such that

$$\hat{\rho}\frac{D\hat{\mathbf{U}}}{D\hat{t}} = -\nabla\hat{p} - \underbrace{\frac{\mathbf{g}\beta(\Delta T)_0 \mathcal{L}}{U_\infty^2}}_{\frac{Gr}{Re^2}}\hat{\rho}\hat{T} + \underbrace{\frac{\mu}{\rho_\infty U_\infty \mathcal{L}}}_{\frac{1}{Re}}\nabla^2\hat{\mathbf{U}}, \qquad (7.25)$$

where $Gr = \frac{|\mathbf{g}||\beta|(\Delta T)_0|\mathcal{L}^3}{\nu^2}$ [–] is known as the Grashof number, which characterizes the relative strength of buoyant to viscous forces [5]. Again, the second term on the right hand side is in the opposite direction of the body acceleration $\mathbf{g}$ [ms^{-2}], assuming that $(\Delta T)_0 > 0$ and $T > T_\infty$ [K]. Grashof number is typically calculated by considering the absolute values of body acceleration and temperature differences, so that it is always a positive quantity.

The similitude analysis above indicates that the solutions of the above system of equations for velocity and temperature fields depend on four dimensionless groups. These are the Reynolds number Re [–], Prandtl number Pr [–], Grashof number Gr [–], and Eckert number Ec [–].

7.6 Boundary Layer Equations for Compressible Flows

We apply Reynolds averaging to the transport equations associated with compressible boundary layer flows. In the presence of turbulence, the temperature field can also be decomposed into the mean and fluctuation components

$$T = \langle T\rangle + t. \qquad (7.26)$$

Such temperature fluctuations give rise to an additional heat flux, analogous to the flux of momentum caused by velocity fluctuations. Thus, the total advective heat flux, per unit area, can be expressed in each of the coordinate directions considering this additional flux. For instance, in the x direction, we can write

$$\frac{Q_x}{dA} = \underbrace{\rho c_p \langle U\rangle\langle T\rangle}_{\langle q_x\rangle} + \underbrace{\rho c_p \langle ut\rangle}_{q_x}, \qquad (7.27)$$

where $\langle q_x\rangle$ [Wm^{-2}] represents the mean advective heat flux and q_x [Wm^{-2}] represents the turbulent advective flux in the x direction. There will be more complications for compressible flow since the presence of temperature and pressure fluctuations produces density fluctuations. So we also need to consider this

$$\rho = \langle \rho\rangle + \varrho. \qquad (7.28)$$

In fact the three fluctuations are related according to the ideal gas law. To the first approximation, we can write

$$\frac{p}{\langle P \rangle} = \frac{t}{\langle T \rangle} + \frac{\varrho}{\langle \rho \rangle} \, . \tag{7.29}$$

On top of the turbulent transfer of heat discussed above, the presence of density fluctuations contributes to a new phenomenon in compressible flows. This phenomenon can be expressed by considering an extended version of the normal and shear stresses given in Eq. 7.5. For selected components of the normal and shear stresses we can write

$$\sigma_x = -\langle \rho \rangle \langle u^2 \rangle - 2\langle U \rangle \langle \varrho u \rangle - \underbrace{\langle \varrho u^2 \rangle}_{\approx 0}$$

$$\tau_{xy} = -\langle \rho \rangle \langle uv \rangle - \langle U \rangle \langle \varrho v \rangle - \langle V \rangle \langle \varrho u \rangle - \underbrace{\langle \varrho uv \rangle}_{\approx 0}$$

$$\tau_{xz} = -\langle \rho \rangle \langle uw \rangle - \langle U \rangle \langle \varrho w \rangle - \langle W \rangle \langle \varrho u \rangle - \underbrace{\langle \varrho uw \rangle}_{\approx 0} \, , \tag{7.30}$$

where the third-order turbulence statistics have been neglected in comparison to the second-order turbulence statistics. The terms $\langle \varrho u \rangle$, $\langle \varrho v \rangle$, and $\langle \varrho w \rangle$ [kgm^{-2}s^{-1}] contribute to turbulent flux of mass in the x, y, and z directions, respectively.

The continuity equation by Reynolds averaging will transform Eq. 7.1 into a new form, in which the storage term vanishes for steady flows and

$$\nabla.(\langle \rho \rangle \langle \mathbf{U} \rangle) + \nabla.(\langle \varrho \mathbf{u} \rangle) = 0$$

$$\frac{\partial(\langle \rho \rangle \langle U \rangle)}{\partial x} + \frac{\partial(\langle \rho \rangle \langle V \rangle)}{\partial y} + \frac{\partial(\langle \rho \rangle \langle W \rangle)}{\partial z} + \frac{\partial \langle \varrho u \rangle}{\partial x} + \frac{\partial \langle \varrho v \rangle}{\partial y} + \frac{\partial \langle \varrho w \rangle}{\partial z} = 0 \, . \tag{7.31}$$

For two-dimensional boundary layers, with flow along the x direction and y direction normal to the wall, we can make some simplifications. The terms in the continuity equation involving the span-wise component z can be omitted. Furthermore, we realize that $\langle V \rangle \ll \langle U \rangle$ [ms^{-1}]. In consideration of boundary layer equations, it is conventional to neglect the normal stresses. In addition, it can be assumed that $\langle V \rangle \langle \varrho u \rangle \ll \langle U \rangle \langle \varrho v \rangle$ [kgm^{-1}s^{-2}] and $\frac{\partial \langle \varrho u \rangle}{\partial x} \ll \frac{\partial \langle \varrho v \rangle}{\partial y}$ [kgm^{-3}s^{-1}], so the continuity equation for two-dimensional boundary layer transforms into

$$\frac{\partial(\langle \rho \rangle \langle U \rangle)}{\partial x} + \underbrace{\frac{\partial(\langle \rho \rangle \langle V \rangle)}{\partial y} + \frac{\partial \langle \varrho v \rangle}{\partial y}}_{\equiv \frac{\partial \langle \rho V \rangle}{\partial y}} = 0 \, . \tag{7.32}$$

To derive the boundary layer momentum equation, the general compressible flow momentum Eq. 7.17 can be transformed using the same approximations and neglecting the same terms. For a steady flow, the momentum equation in the x direction can be given as

$$\underbrace{\langle\rho\rangle\langle U\rangle\frac{\partial\langle U\rangle}{\partial x} + \underbrace{(\langle\rho\rangle\langle V\rangle + \langle\varrho v\rangle)}_{\equiv\langle\rho V\rangle}\frac{\partial\langle U\rangle}{\partial y}}_{\text{Advection}} = \underbrace{-\frac{\partial\langle p\rangle}{\partial x}}_{\text{Surface Normal Force}}$$

$$+\underbrace{\frac{\partial}{\partial y}\left(\mu\frac{\partial\langle U\rangle}{\partial y}\right)\underbrace{-\frac{\partial(\langle\rho\rangle\langle uv\rangle)}{\partial y}}_{\frac{\partial\tau_{xy}}{\partial y}}}_{\text{Surface Shear Force}} .$$

$$(7.33)$$

It is noted that $\langle\rho V\rangle \equiv \langle\rho\rangle\langle V\rangle + \langle\varrho v\rangle$ [kgm^{-2}s^{-1}] is defined as the total mass flux in the y direction. Furthermore, term $\tau_{xy} = -\langle\rho\rangle\langle uv\rangle$ [kgm^{-1}s^{-2}] appears in the equation, which can be considered as the apparent shear stress. It is interesting that we have left the transport equation for momentum in the y direction, i.e. $\langle V\rangle$ [ms^{-1}], undetermined.

To derive the boundary layer energy equation, the general compressible flow energy Eq. 7.12 can be transformed using the same approximations and neglecting the same terms. For steady flow, the energy equation can be written as

$$\underbrace{c_p\left(\langle\rho\rangle\langle U\rangle\frac{\partial\langle T\rangle}{\partial x} + \langle\rho V\rangle\frac{\partial\langle T\rangle}{\partial y}\right)}_{\text{Advection}} = \underbrace{\frac{\partial}{\partial y}\left(k\frac{\partial\langle T\rangle}{\partial y}\right)}_{\text{Diffusion}} \underbrace{-\frac{\partial q_y}{\partial y}}_{\text{Turbulent Heat Flux Divergence}}$$

$$+ \underbrace{\langle\mu\Phi\rangle}_{\text{Dissipation}} + \underbrace{\langle U\rangle\frac{\partial\langle p\rangle}{\partial x}}_{\text{Pressure}} , \qquad (7.34)$$

where $q_y = c_p\langle\rho\rangle\langle vt\rangle$. The term $\langle\mu\Phi\rangle$ [Wm^{-3}] accounts for the mean value of the dissipation, for which the following approximation can be given

$$\langle\mu\Phi\rangle = \left(\mu\frac{\partial\langle U\rangle}{\partial y} + \tau_{xy}\right)\frac{\partial\langle U\rangle}{\partial y} . \qquad (7.35)$$

Finally, the equation of ideal gas is another equation that pertains to boundary layer compressible turbulent flows. The ideal gas law can be given as

$$\langle p\rangle = \langle\rho\rangle R\langle T\rangle . \qquad (7.36)$$

In boundary layer compressible turbulent flow approximation, we may replace the absolute pressure with the modified pressure $p = P$ [Pa], which is perfectly legitimate if effects of gravitational body forces are to be neglected. This is certainly the case in many problems involving airflow around airfoils and other similar cases.

So, all together, the boundary layer compressible turbulent flows require 4 simultaneous equations to solve for four variables: $\langle U \rangle$ [ms^{-1}], $\langle p \rangle$ [Pa], $\langle \rho \rangle$ [kgm^{-3}], and $\langle T \rangle$ [K]. The corresponding equations available are the continuity Eq. 7.32 (1), momentum Eq. 7.33 (1), the energy Eq. 7.34 (1), and the ideal gas law 7.36.

7.7 Mach Number

The Mach number characterizes the behaviour of high-speed compressible flows. It is defined as the ratio of the fluid speed to that of the local speed of sound

$$M = \frac{S}{C} , \qquad (7.37)$$

where S [ms^{-1}] is the flow speed and C [ms^{-1}] is the local speed of sound. Generally, the compressibility effects are accentuated with higher Mach numbers. Figure 7.2 shows the flow regimes associated with the Mach number, which is typically used in the analysis and design of aircraft. Flows with Mach numbers lower than about 0.2 are considered as incompressible flows, with negligible density

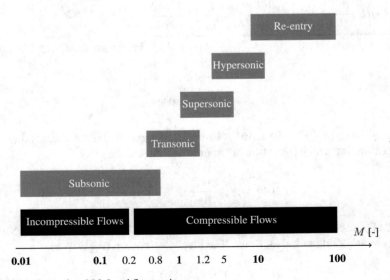

Fig. 7.2 Mach number M [–] and flow regimes

variations, while those with higher Mach numbers are considered as compressible flows, for which density, pressure, and temperature changes must be accounted for. In the subsonic regime $M < 0.8$ [–], all flow speeds around the airfoil of an aircraft are subsonic. In the transonic regime $0.8 < M < 1.2$ [–], some areas of the airfoil experience supersonic flows, but other areas still experience subsonic flows. However, in the supersonic regime $1.2 < M < 5$ [–], most areas of the airfoil experience supersonic flows, except possibly a few stagnation points. In the hypersonic $5 < M < 10$ [–] and re-entry $10 < M$ [–] regimes, extreme temperatures or plasma may be experienced in the flow, which complicates aircraft design. In addition to temperature and plasma effects, aircraft should be designed to withstand shock waves produced in supersonic flows [2].

Problems

7.1 The ideal gas law is given as $P = \rho R T$ [Pa], where P [Pa] is the pressure, ρ [kgm^{-3}] is the density, R [Jkg^{-1}K^{-1}] is the specific gas constant, and T [K] is the temperature. Using Reynolds decomposition pressure, density, and temperature can be written as $P = \langle P \rangle + p$ [Pa], $\rho = \langle \rho \rangle + \varrho$ [kgm^{-3}], and $T = \langle T \rangle + t$ [K], respectively. Substitute these compositions into the ideal gas law, and by ignoring the terms involving products of fluctuations, show that

$$\frac{p}{\langle P \rangle} = \frac{t}{\langle T \rangle} + \frac{\varrho}{\langle \rho \rangle} . \tag{7.38}$$

7.2 Show that if ρ [kgm^{-3}] is a scalar and $\mathbf{U}$ [ms^{-1}] is a vector, then the following identity is valid

$$\nabla.(\rho \mathbf{U}) = \mathbf{U}.\nabla \rho + \rho \nabla.\mathbf{U} . \tag{7.39}$$

7.3 Using variable substitutions, derive the non-dimensional forms of the transport equations for compressible flow provided by similitude analysis.

References

1. Acheson D J (1990) Elementary fluid dynamics. Oxford University Press, Oxford
2. White F M (2003) Fluid mechanics. McGraw-Hill Higher Education, New York
3. Schlichting H (1979) Boundary-layer theory, 7th edn. McGraw-Hill, New York
4. Ettema R (2000) Hydraulic modelling: concepts and practice. American Society of Civil Engineering (ASCE), Reston
5. Aliabadi A A (2013) Dispersion of expiratory airborne droplets in a model single patient hospital recovery room with stratified ventilation. University of British Columbia, Vancouver

Chapter 8
Scales of Turbulent Motion

Abstract This chapter discusses the scales of turbulent motion. The energy cascade and the Kolmogorov hypotheses are introduced. The scales of turbulent motion are grouped under the energy-containing and the universal equilibrium ranges. The universal equilibrium range is further subdivided into inertial and dissipation subranges. The concept of energy spectrum is introduced, and its links to the turbulence kinetic energy and dissipation rate are established. The two-point correlation function and its relation to the integral lengthscale are specified. The structure function and its relation to the dissipation rate are specified. Finally, the Taylor hypothesis, also known as the frozen eddy hypothesis, is introduced, which provides an important simplification made in the analysis of many environmental flows.

8.1 Preliminaries

The methodology developed in this chapter is adapted from [1]. The study of scales of turbulent motion revolves around the understanding of various time, length, and velocity scales at which flow instabilities form and progress into turbulent motion of different scales, i.e. different time, length, and velocity scales. This hints that turbulent motion is intrinsically transient and moves towards various states governed by known hypotheses.

The *energy cascade* hypotheses were first introduced by Lewis Fry Richardson (1881–1953) and referred to in Chap. 1. According to this hypothesis, turbulence kinetic energy (Eq. 4.11) is produced and enters a turbulent flow at the largest lengthscales of motion. This energy is then transferred by inviscid processes to smaller and smaller scales until, at the smallest scales, the energy is dissipated by viscous action. In a famous poem Richardson describes this hypothesis:

> Big whirls have little whirls that feed on their velocity, and little whirls have lesser whirls and so on to viscosity.

Andrey Nikolaevich Kolmogorov (1903–1987) later quantified the energy cascade hypothesis using mathematical formulations. Kolmogorov performed this calculation by other refined hypotheses known as the *Kolmogorov hypotheses*.

8.2   The Energy Cascade and Kolmogorov Hypotheses

Consider a turbulent flow with characteristic velocity scale $\mathcal{U}$ [m s^{-1}] and characteristic lengthscale $\mathcal{L}$ [m] and a Reynolds number $Re = \mathcal{U}\mathcal{L}/\nu$ [–] that is very high. Richardson viewed the turbulent flow to be composed of *eddies* of different sizes. Eddies of size ℓ [m] have a characteristic velocity $u(\ell)$ [m s^{-1}] and timescale $\tau(\ell) \equiv \ell/u(\ell)$ [s]. Each eddy is conceived to be a turbulent motion, localized within a region of size ℓ [m] that has a coherent structure. For instance, it can be a blob of fluid spinning.

The largest eddies that are formed in turbulent flow are designated with subscript 0, i.e. ℓ_0 [m], and are comparable to the flow lengthscale $\mathcal{L}$ [m]. Likewise, the velocities associated with these eddies, i.e. $u_0 \equiv u(\ell_0)$ [m s^{-1}], are comparable to the flow velocity scale $\mathcal{U}$ [m s^{-1}]. These velocities are in the order of velocity fluctuations and related to the root mean square turbulence kinetic energy, i.e. $u' \equiv (\frac{2}{3}k)^{1/2}$ [m s^{-1}]. It is possible to associate a Reynolds number to these eddies $Re_0 = u_0\ell_0/\nu$ [–], which is also comparable to the flow Reynolds number Re [–]. In this scale the effect of viscosity is negligible since the flow is dominated by inertial effects.

The larger eddies continue to break up into smaller and smaller eddies, each characterized by a Reynolds number $Re(\ell) = u(\ell)\ell/\nu$ [–]. When this Reynolds number is sufficiently small, at some point the viscous effects dominate and the kinetic energy dissipates.

The rate at which turbulence kinetic energy dissipates by viscous effects is called the *dissipation rate* ϵ [m^{2}s^{-3}]. In statistically stationary turbulent flow the *energy transfer rate* $\mathcal{T}_{EI}$ [m^{2}s^{-3}] at each eddy lengthscale is the same, which provides a convenient method for the calculation of the dissipation rate. At the beginning of the process, the eddies have energy in the order of u_0^2 [m^{2}s^{-2}] and timescale in the order of $\tau_0 = \ell_0/u_0$ [s], so the rate of transfer of energy and ultimately the dissipation rate are in the order of $u_0^2/\tau_0 = u_0^3/\ell_0$ [m^{2}s^{-3}]. An interesting result is that this rate is independent of flow kinematic viscosity ν [m^{2}s^{-1}], provided that the Reynolds number of the flow is sufficiently high.

Figure 8.1 provides an analogy for the rate at which turbulence kinetic energy is transferred via the energy cascade from the energy-containing range at lengthscale ℓ_0 [m] all the way down to the dissipation subrange at the smallest scale η [m]. The analogy demonstrates the turbulence kinetic energy as mass of water flowing down a water fall via a series of buckets. Note that mass and energy should not be

Statistically stationary turbulent flow

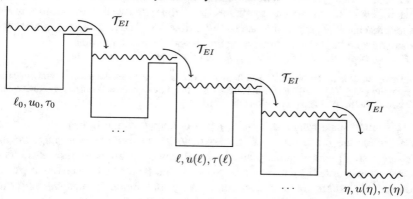

Statistically non-stationary turbulent flow

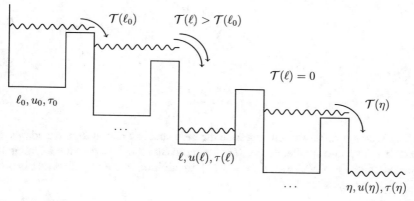

Fig. 8.1 The analogy of water fall via a series of buckets to explain the concept of energy cascade under statistically stationary and non-stationary conditions

confused in this analogy. Under the statistically stationary turbulent flow conditions, the rate of energy transfer is the same throughout the energy cascade. However, under the statistically non-stationary conditions, the energy transfer rate may be higher or lower at each stage of the cascade.

It is now time to introduce *Kolmogorov hypotheses* [2] to describe and quantify the energy cascade hypothesis:

Kolmogorov hypothesis of local isotropy states that:

At sufficiently high Reynolds number, the small-scale turbulent motions ($\ell \ll \ell_0$) [m] are statistically isotropic.

In other words, at the isotropic scales, the directional information of the large-scale eddies is lost as the energy passes down the cascade. It is useful to introduce the lengthscale ℓ_{EI} [m], which is usually a fraction of ℓ_0 [m] as the scale at which the anisotropic larger eddies break up into isotropic smaller eddies.

Kolmogorov's first similarity hypothesis states that:

In every turbulent flow at sufficiently high Reynolds number, the statistics of the small-scale motions ($\ell \leq \ell_{EI}$) [m] have a universal form that is uniquely determined by kinematic viscosity ν [m^{2}s^{-1}] and dissipation rate ϵ [m^{2}s^{-3}].

This lengthscale range of $\ell \leq \ell_{EI}$ [m] is known as the *universal equilibrium range*, while the range $\ell_{EI} \leq \ell$ [m] is known as the *energy-containing range*. In this range small eddies quickly adapt to maintain a dynamic equilibrium with the energy transfer rate $\mathcal{T}_{EI}$ [m^{2}s^{-3}] imposed by the larger eddies. Given ν [m^{2}s^{-1}] and ϵ [m^{2}s^{-3}], there are unique length, velocity, and timescales that can be formed. These are known as the *Kolmogorov scales*:

$$\eta \equiv \left(\frac{\nu^3}{\epsilon}\right)^{1/4}, \tag{8.1}$$

$$u_\eta \equiv (\epsilon\nu)^{1/4}, \tag{8.2}$$

$$\tau_\eta \equiv \left(\frac{\nu}{\epsilon}\right)^{1/2}. \tag{8.3}$$

The Kolmogorov scales characterize the very smallest and dissipative eddies that do not break up into smaller eddies. The Reynolds number based on the Kolmogorov scales is unity, i.e. $Re(\eta) = \eta u_\eta/\nu = 1$ [–]. In addition, the dissipation rate is given by

$$\epsilon = \nu \left(\frac{u_\eta}{\eta}\right)^2 = \frac{\nu}{\tau_\eta^2}. \tag{8.4}$$

On the smallest scales in high Reynolds number turbulent flows, turbulent velocity fields are statistically similar, that is they are statistically identical when they are scaled by the Kolmogorov scales. The ratios of the smallest to largest scales are given by

$$\frac{\eta}{\ell_0} \sim Re^{-3/4}, \tag{8.5}$$

$$\frac{u_\eta}{u_0} \sim Re^{-1/4}, \tag{8.6}$$

$$\frac{\tau_\eta}{\tau_0} \sim Re^{-1/2}. \tag{8.7}$$

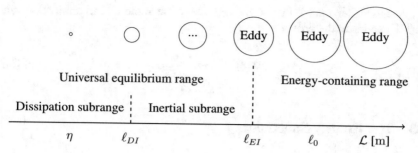

Fig. 8.2 Eddy size ℓ [m] and various ranges of energy cascade and eddy lengthscales for very high Reynolds number flows

These relationships are extremely useful because without having to even measure or simulate turbulent flow, it is possible to infer the Kolmogorov scales having the characteristic length, velocity, and timescales of the flow that are in the order of the largest eddy scales as well.

Kolmogorov's second similarity hypothesis states that:

> In every turbulent flow at sufficiently high Reynolds number, the statistics of the motions of scale ℓ in the range $\eta \ll \ell \ll \ell_0$ [m] have a universal form that is uniquely determined by ϵ [m²s⁻³] independent of ν [m²s⁻¹].

This hypothesis builds on the previous hypotheses in a way that if ℓ [m] is significantly larger than the Kolmogorov lengthscale η [m], then only inertial effects are dominant and the viscous effects should not determine the statistics of the turbulent motions. For this hypothesis, it is useful to define another lengthscale ℓ_{DI} [m] that is a significant multiple of η [m]. We can then specify a range for ℓ [m] such as $\ell_{DI} \leq \ell \leq \ell_{EI}$ [m] known as the *inertial subrange* and another range for ℓ [m] such as $\eta \leq \ell \leq \ell_{DI}$ [m] known as the *dissipation subrange*. Figure 8.2 demonstrates the range of eddy lengthscales and the energy cascade ranges.

For the inertial subrange, having ℓ [m] and ϵ [m²s⁻³], it is possible to estimate characteristic velocity and timescales for an eddy lengthscale of ℓ [m] as follows:

$$u(\ell) = (\epsilon \ell)^{1/3}, \tag{8.8}$$

$$\tau(\ell) = (\ell^2/\epsilon)^{1/3}. \tag{8.9}$$

It is worth reiterating that for statistically stationary flow, and in the inertial subrange, the rate of energy transfer $\mathcal{T}(\ell)$ [m²s⁻³] from eddies larger than ℓ [m]

to eddies smaller than ℓ [m] is approximately the same as the dissipation rate ϵ [m^{2}s^{-3}] and is in the order of

$$\mathcal{T}(\ell) \sim \epsilon \sim \frac{u(\ell)^2}{\tau(\ell)}. \tag{8.10}$$

8.3 The Energy Spectrum

The turbulence kinetic energy in a flow is distributed among different eddies with different lengthscales (and the associated time and velocity scales). This distribution can be quantified by the *energy spectrum function* $E(\kappa)$ [m^{3}s^{-2}], where $\kappa = 2\pi/\ell$ [m^{-1} or Radm^{-1}] is the wave number associated with each eddy size ℓ [m]. The turbulence kinetic energy $k_{(\kappa_a,\kappa_b)}$ [m^{2}s^{-2}] in the wave number range (κ_a, κ_b) [m^{-1}] can be given by

$$k_{(\kappa_a,\kappa_b)} = \int_{\kappa_a}^{\kappa_b} E(\kappa)d\kappa. \tag{8.11}$$

A similar notion was introduced in Chap. 3 with Eq. 3.24 for the frequency spectrum function. In that case, it must be cautioned that the unit for the energy spectrum function of frequency, i.e. $E(\omega)$ [m^{2}s^{-1}], was different. Likewise, the dissipation rate ϵ [m^{2}s^{-3}] for the same range of wave numbers can be given by

$$\epsilon_{(\kappa_a,\kappa_b)} = \int_{\kappa_a}^{\kappa_b} 2\nu\kappa^2 E(\kappa)d\kappa. \tag{8.12}$$

For homogeneous turbulence and in the inertial subrange $(\kappa_{EI} < \kappa < \kappa_{DI})$ [m^{-1}] the energy spectrum can be given by well-defined functions. One example is

$$E(\kappa) = C\epsilon^{2/3}\kappa^{-5/3}, \tag{8.13}$$

where C [–] is a constant. This is also known as the Kolmogorov $-5/3$ spectrum and can successfully describe the energy spectrum for many turbulent flows. Figure 8.3 shows the normalized energy spectrum function versus the wave number for high frequency measurement of wind velocity vector above an urban canyon [3]. The Kolmogorov spectrum indicates that the $-5/3$ slope is present for two orders of magnitude of the wave number.

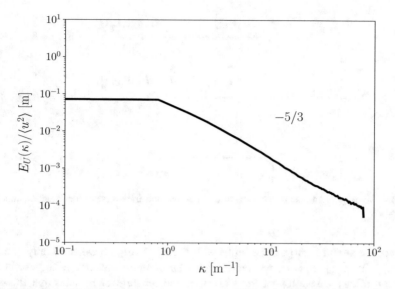

Fig. 8.3 Example of a normalized energy spectrum function versus the wave number that exhibits the Kolmogorov $-5/3$ spectrum; data collected using ultrasonic anemometers to measure the horizontal component of wind velocity vector above an urban canyon at the University of Guelph, Canada [3]

8.4 Two-Point Correlation

The autocorrelation function is defined in Chap. 3 as Eq. 3.18 for statistically stationary turbulence as a one-point and two-time statistic, i.e. it quantified the correlation of velocity fluctuations at one point in the flow but with a variable time shift. Here we redefine the autocorrelation function as a two-point and one-time statistic given by

$$R_{ij}(\mathbf{r}, \mathbf{x}, t) \equiv \langle u_i(\mathbf{x} + \mathbf{r}, t) u_j(\mathbf{x}, t) \rangle, \tag{8.14}$$

which was also introduced in Chap. 3 by Eq. 3.32. If such a correlation sharply decreases with the increasing distance between the two points, then turbulent flow ought to exhibit small turbulent eddies. On the other hand if this correlation is significant even at larger distances between the two points, then the flow exhibits large turbulent eddies. In the limit of no separation distance between the two points, the correlation represents the Reynolds stresses. In the limit of a very large distance between the two points, the correlation ought to drop to zero since a flow cannot exhibit infinitely large turbulent eddies [4].

Some simplifications can be made. For homogeneous and isotropic turbulence, introduced in Chap. 3, the dependence on $\mathbf{x}$ [m] can be removed, i.e. only $R_{ij}(\mathbf{r}, t)$ [m^{2}s^{-2}] shall be considered. Further, since the coordinate system can be freely

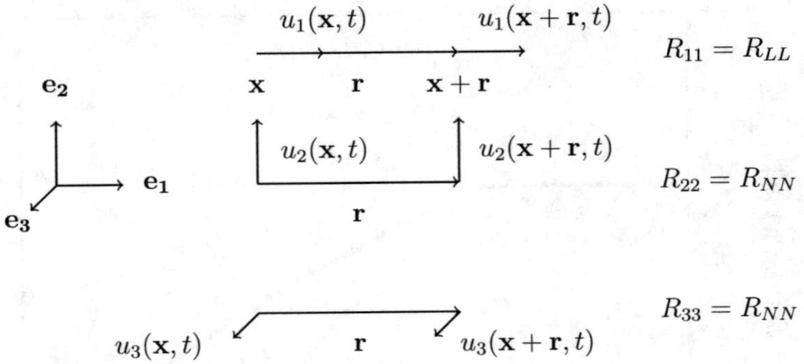

Fig. 8.4 Velocity components involved in longitudinal and transverse autocorrelation functions for $\mathbf{r} = \mathbf{e}_1 r$

rotated, the vector $\mathbf{r}$ [m] can be replaced by scalar r [m], so we shall only consider $R_{ij}(r, t)$ [m^2s^{-2}]. It is convenient to express components of the autocorrelation function after the rotation of the Cartesian coordinate system such that the coordinate axis $\mathbf{e}_1$ is aligned with vector $\mathbf{r}$. Note that this is perfectly allowed for isotropic turbulence. The other two coordinate axes $\mathbf{e}_2$ and $\mathbf{e}_3$ then become normal or perpendicular to $\mathbf{r}$. This is depicted in Fig. 8.4.

After the rotation of the Cartesian coordinate system, the components of the autocorrelation function are $R_{11} = R_{LL}$ and $R_{22} = R_{33} = R_{NN}$ [m^2s^{-2}] that are known as *longitudinal autocorrelation function* and *transverse autocorrelation function*, respectively. Note that the isotropic condition requires $R_{22} = R_{33}$ [m^2s^{-2}], but there is no reason for the longitudinal and transverse autocorrelation functions to be the same.

It was stated earlier that if the separation distance between the two points is zero, then the autocorrelation reduces to Reynolds stresses so that

$$R_{ij}(0, t) = \langle u_i u_j \rangle. \tag{8.15}$$

Due to isotropic condition, this implies that $\langle u^2 \rangle \equiv \langle u_1^2 \rangle = \langle u_2^2 \rangle = \langle u_3^2 \rangle$ [m^2s^{-2}]. The autocorrelation functions can be expressed in the general forms

$$R_{11}(r, t) = R_{LL}(r, t) = \langle u^2 \rangle f(r, t), \tag{8.16}$$

$$R_{22}(r, t) = R_{33}(r, t) = R_{NN}(r, t) = \langle u^2 \rangle g(r, t), \tag{8.17}$$

where $f(0, t) = g(0, t) = 1$ [–] by definition. Figure 8.5 shows the well-behaved computed normalized autocorrelation functions measured by an aircraft [5]. In this figure, longitudinal and transverse autocorrelation functions are shown by $f(r, t) = R_{uu}$ and $g(r, t) = R_{ww}$ [–], respectively. As will be shown later in this chapter, using the Taylor hypothesis, the time shift axis t [s] can be related to the spatial shift axis r [m], where $r = Vt$ [m], in which V [ms^{-1}] is the average speed of the

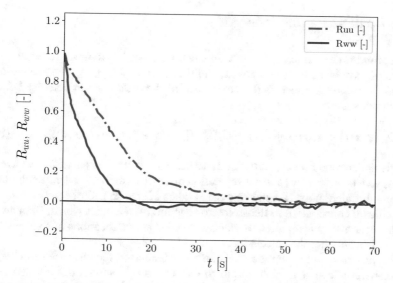

Fig. 8.5 Example of normalized autocorrelation functions measured by an aircraft [5]

aircraft. Autocorrelation functions are computed without high-pass filtering of input data to ensure that all scales of fluctuations are accounted for. From the figure, it can be seen that the normalized R_{ww} and R_{uu} [–] start near unity and decrease to near zero by about $t = 10\,\mathrm{s}$ and $40\,\mathrm{s}$, respectively.

It is possible to calculate *integral lengthscales* if the autocorrelation functions are available. The first of the lengthscales obtained from $f(r, t)$ [–] is the *longitudinal integral lengthscale* given by

$$L_{11}(t) = L_{LL}(t) \equiv \int_0^\infty f(r, t)\,dr, \qquad (8.18)$$

which is the characteristic of larger eddies. In isotropic turbulence, the *transverse integral lengthscale* is obtained from $g(r, t)$ [–] given by

$$L_{22}(t) = L_{33}(t) = L_{NN}(t) \equiv \int_0^\infty g(r, t)\,dr, \qquad (8.19)$$

which is the characteristic of smaller eddies. If desired, an average of the two eddy lengthscales can be used to report one eddy lengthscale for the turbulent flow of interest.

8.5   Structure Functions

The second-order velocity structure functions are useful statistics to describe the nature of turbulent flows, particularly the dissipation rate [6, 7]. The second-order *velocity structure function* is the covariance of the difference in velocity between two points $\mathbf{x}$ and $\mathbf{x} + \mathbf{r}$ [m]

$$D_{ij}(\mathbf{r}, \mathbf{x}, t) \equiv \langle [U_i(\mathbf{x} + \mathbf{r}, t) - U_i(\mathbf{x}, t)][U_j(\mathbf{x} + \mathbf{r}, t) - U_j(\mathbf{x}, t)] \rangle. \qquad (8.20)$$

It is understood that the structure function is computed as an ensemble average for a given location $\mathbf{x}$ [m] and the separation distance $\mathbf{r}$ [m], and at a particular time t [s] for a turbulent flow. It seems that only eddies of size $|\mathbf{r}|$ [m] or smaller can make a significant contribution to the structure function [4]. In other words, if an eddy is much larger than $|\mathbf{r}|$ [m], then it affects velocities at two locations in a similar way, which does not alter the correlation.

We can make some simplifications. In homogeneous turbulence the velocity fluctuations are independent of location, so the structure function does not depend on location $\mathbf{x}$ [m], i.e. we only have $D_{ij}(\mathbf{r}, t)$ [m^{2}s^{-2}]. Using Kolmogorov hypothesis of local isotropy for $\mathbf{r} \ll \mathcal{L}$ [m], we can consider $D_{ij}(\mathbf{r}, t)$ [m^{2}s^{-2}] as an isotropic function of $\mathbf{r}$ [m]. It is convenient to express components of the structure function after the rotation of the Cartesian coordinate system such that the coordinate axis $\mathbf{e}_1$ is aligned with vector $\mathbf{r}$ [m]. Note that this is perfectly allowed for isotropic turbulence. The other two coordinate axes $\mathbf{e}_2$ and $\mathbf{e}_3$ then become normal or perpendicular to $\mathbf{r}$ [m]. This is depicted in Fig. 8.6.

After the rotation of the Cartesian coordinate system, the components of the velocity structure function are $D_{11} = D_{LL}$ and $D_{22} = D_{33} = D_{NN}$ [m^{2}s^{-2}] that are known as *longitudinal structure function* and *transverse structure function*, respectively. Note that the isotropic condition requires $D_{22} = D_{33}$ [m^{2}s^{-2}], but

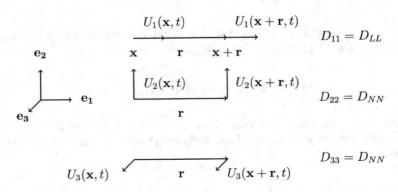

Fig. 8.6 Velocity components involved in longitudinal and transverse structure functions for $\mathbf{r} = \mathbf{e}_1 r$

there is no reason for the longitudinal and transverse structure functions to be the same.

Another simplification that can be made is that in isotropic turbulence D_{LL} and D_{NN} [m^{2}s^{-2}] are independent of the direction of the vector $\mathbf{r}$ [m], so that the vector quantity $\mathbf{r}$ [m] can be replaced by scalar r [m]. Given the first Kolmogorov hypothesis, when $r \ll \mathcal{L}$ [m], D_{ij} [m^{2}s^{-2}] can be uniquely described by ν [m^{2}s^{-1}] and ϵ [m^{2}s^{-3}] in the universal equilibrium range. Further, when $\eta \ll r \ll \mathcal{L}$ [m], D_{ij} [m^{2}s^{-2}] can be uniquely described by ϵ [m^{2}s^{-3}] in the inertial subrange. For the inertial subrange and statistically homogeneous and isotropic conditions, it has been shown that

$$D_{LL}(r, t) = C_2(\epsilon r)^{2/3}, \tag{8.21}$$

$$D_{NN}(r, t) = \frac{4}{3}C_2(\epsilon r)^{2/3}, \tag{8.22}$$

where $C_2 \approx 2.0$ [–] is a universal constant.

8.6 Taylor Hypothesis

Measurement of the one-time and two-point velocity correlation function $R_{ij}(\mathbf{r})$ [m^{2}s^{-2}] experimentally is very difficult since this requires two stationary probes that need to be used in a large number of experiments with variable separation distance $\mathbf{r}$ [m]. Alternatively, it is possible to use a moving probe in turbulent flow that travels at very high speed V [m s^{-1}] along $\mathbf{r}$ [m]. Let us assume that $\mathbf{r}$ [m] is in the direction of the first component of the Cartesian coordinate system, i.e. $\mathbf{r} = r\mathbf{e}_1$ [m], where $\mathbf{e}_1$ is the unit vector along the first component. Now if the autocorrelation for the moving probe with a time shift s [s] is calculated, we obtain $R_{ij}^{(m)}(s)$ [m^{2}s^{-2}], where the superscript (m) signifies the moving probe. On the other hand, for a time shift of s [s], the probe has moved equal to

$$r = Vs. \tag{8.23}$$

It is possible to show that if the probe is moving infinitely fast in the flow, then this autocorrelation is equal to the one-time and two-point velocity correlation, i.e.

$$R_{ij}^{(m)}(s) = R_{ij}(r). \tag{8.24}$$

This concept is the basis of turbulence measurements using flying hot wire anemometers, where a hot wire anemometer that is very sensitive to flow measurements with a large sampling rate is flown in a turbulent flow to reveal one-time and two-point velocity correlation functions [8]. In other atmospheric measurement studies using aircraft, the same approximation can be made given the fact that aircraft velocity is usually very fast so that the timescale involved in flying through turbulence structures is significantly shorter compared to the timescale of atmospheric eddies, at least eddies in the inertial subrange or energy-containing range [5, 9, 10].

In many studies, particularly in atmospheric flows, there is a simpler approach to use. In this technique, only a single stationary probe is required for the approximation to be valid so that

$$R_{ij}(s) = R_{ij}(r),$$
(8.25)

where now $V = r/s$ [m s^{-1}] represents the average wind speed. The approximation of spatial correlations by temporal correlations is known as the *Taylor hypothesis* [11] and is only valid for the *frozen turbulence approximation*. This approximation states that eddies can be conceived as frozen and moving with the flow as they travel past a stationary probe. This condition occurs in the atmosphere when turbulent fluctuations are much smaller in magnitude than the average flow velocity, for instance

$$\frac{u_1}{\langle U_1 \rangle} \ll 1.$$
(8.26)

Problems

8.1 Show that at Kolmogorov scales the Reynolds number of the flow is unity, i.e.

$$Re(\eta) = \frac{\eta u_\eta}{\nu} = 1.$$
(8.27)

8.2 Show that the ratios of the smallest Kolmogorov to the largest scales of turbulence in a flow are given by

$$\frac{\eta}{\ell_0} \sim Re^{-3/4},$$
(8.28)

$$\frac{u_\eta}{u_0} \sim Re^{-1/4},$$
(8.29)

$$\frac{\tau_\eta}{\tau_0} \sim Re^{-1/2}.$$
(8.30)

8.3 For the inertial subrange and very high Reynolds number turbulent flows, show that it is possible to approximate characteristic velocity and timescales for an eddy lengthscale of ℓ [m] as follows:

$$u(\ell) \sim u_0(\ell/\ell_0)^{1/3}, \tag{8.31}$$

$$\tau(\ell) \sim \tau_0(\ell/\ell_0)^{2/3}. \tag{8.32}$$

8.4 For homogeneous and statistically stationary turbulence, the energy spectrum function for the inertial subrange can be given using a power law such that

$$E(\kappa) = A\kappa^{-p}, \tag{8.33}$$

where A and p are the constants. Assuming that $1 < p < 3$ and that the inertial subrange is represented by $0 < \kappa < \infty \, [m^{-1}]$, show that the turbulence kinetic energy and the dissipation rate over specific ranges of wave numbers below can be given using

$$k_{(\kappa,\infty)} = \frac{A}{p-1}\kappa^{-(p-1)}, \tag{8.34}$$

$$\epsilon_{(0,\kappa)} = \frac{2\nu A}{3-p}\kappa^{3-p}. \tag{8.35}$$

8.5 For homogeneous and isotropic turbulence, sketch an approximate plot of the logarithm of velocity structure functions $D_{LL}(r, t)$ and $D_{NN}(r, t)$ [m^{2}s^{-2}] versus r [m]. Identify the zero and the horizontal asymptotic limits for the logarithm of the structure functions. Identify the portion of the plot for which you know the slope of the curves. Identify the dissipation subrange, inertial subrange, and the energy-containing range on the plot. Within the inertial subrange, identify which curve is above the other.

8.6 Show by definition why for isotropic turbulence $f(0, t) = g(0, t) = 1$ [–].

8.7 For a flow with isotropic turbulence conditions, the autocorrelation functions at a particular time t [s] in the longitudinal and transverse directions are given as

$$f(r) = \frac{1}{1+r^2} \text{ and } g(r) = \cos(r)e^{-r}, \tag{8.36}$$

where r [m] is the distance. Show that for this flow and at this particular time, the longitudinal integral lengthscale and the transverse integral lengthscale are given by

$$L_{LL} = \frac{\pi}{2}\text{m and } L_{NN} = \frac{1}{2}\text{m}. \tag{8.37}$$

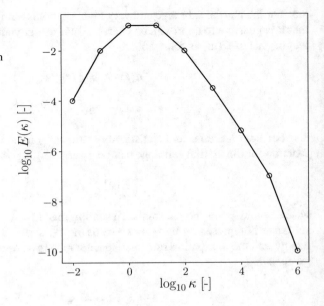

Fig. 8.7 Energy spectrum function: base ten logarithm of energy spectrum function $E(\kappa)$ [m^{3}s^{-2}] versus base ten logarithm of wave number κ [m^{-1}]

8.8 For an experiment, in which flow is homogeneous and statistically stationary, the energy spectrum function is obtained in a few discrete points. Figure 8.7 shows the base ten logarithm of energy spectrum function $E(\kappa)$ [m^{3}s^{-2}] versus the base ten logarithm of wave number κ [m^{-1}].

Provide a range of wave numbers in the energy-containing range, i.e. $\kappa < \kappa_{EI}$ [m^{-1}]; provide a range of wave numbers in the inertial subrange, i.e. $\kappa_{EI} < \kappa < \kappa_{DI}$ [m^{-1}]; and provide a range of wave numbers in the dissipation subrange, i.e. $\kappa > \kappa_{DI}$ [m^{-1}]. The base ten logarithm of wave number is taken for a unitless wave number normalized by unit wave number, i.e. $1\,\mathrm{m}^{-1}$. The base ten logarithm of energy spectrum function is taken for a unitless energy spectrum normalized by unit spectrum $1\,\mathrm{m}^3\,\mathrm{s}^{-2}$.

8.9 Show that the unit for energy spectrum function $E(\kappa)$ is [m^3 s^{-2}].

8.10 A mathematician is challenged with the following question. For a particular airflow system, the energy spectrum function in a specific range of wave numbers κ [m^{-1}] within the inertial subrange is given as

$$E(\kappa) = e^{-\kappa}, \tag{8.38}$$

in units of [m^{3}s^{-2}]. Assuming that the kinematic viscosity of air is $\nu = 1.5 \times 10^{-5}$ m^{2}s^{-1}, help her calculate the turbulence kinetic energy k [m^{2}s^{-2}] and the turbulence kinetic energy dissipation rate ϵ [m^{2}s^{-3}] contained in the range of wave numbers $[1, 10]\,\mathrm{m}^{-1}$ within the inertial subrange. Hint: both of these quantities can be obtained by integrating a function containing the energy spectrum.

8.11 The model spectrum for energy density provided thus far only describes the inertial subrange. However, other model spectra have been proposed that cover more subranges of the energy cascade for statistically stationary but isotropic or anisotropic turbulence, such as energy-containing, inertial, and dissipation subranges. One such model can be written as

$$E(\kappa) = C\epsilon^{2/3}\kappa^{-5/3}f_L(\kappa L)f_\eta(\kappa\eta), \tag{8.39}$$

where C [–] and L [m] are the constants, ϵ [m^{2}s^{-3}] is the turbulence kinetic energy dissipation rate, κ [m^{-1}] is the wave number, and η [m] is the Kolmogorov lengthscale. Here $f_L(\kappa L)$ [–] and $f_\eta(\kappa\eta)$ [–] are multiplying functions to correct the model spectrum for the energy-containing and dissipation subranges, respectively [1]. $f_L(\kappa L)$ [–] is defined as

$$f_L(\kappa L) = \left(\frac{\kappa L}{[(\kappa L)^2 + c_L]^{1/2}}\right)^{5/3+p_0}, \tag{8.40}$$

where c_L [–] and p_0 [–] are the constants. p_0 [–] is taken as a different constant for anisotropic and isotropic flows [12–14]. $f_\eta(\kappa\eta)$ [–] is defined as

$$f_\eta(\kappa\eta) = \exp\{-\beta\{[(\kappa\eta)^4 + c_\eta^4]^{1/4} - c_\eta\}\}, \tag{8.41}$$

where β [–] and c_η [–] are the constants. (a) Show that in the limit where $\kappa \to 0$ m^{-1}, i.e. when the energy-containing range is considered, $f_\eta(\kappa\eta)$ [–] behaves like a constant, and the model spectrum behaves like

$$E(\kappa) \sim \kappa^{p_0}. \tag{8.42}$$

(b) Assuming $c_\eta = 0$ [–], show that in the limit where $\kappa \to \infty$ m^{-1}, i.e. when the dissipation subrange is considered, $f_L(\kappa L)$ [–] approaches unity, and the model spectrum behaves like

$$E(\kappa) \sim \kappa^{-5/3}\exp(-\beta\kappa\eta). \tag{8.43}$$

Note that since the velocity field is infinitely differentiable, for large κ [m^{-1}], the spectrum function decays more rapidly than any power of κ [m^{-1}], say $-5/3$, so the dissipation subrange is described by the model spectrum appropriately. Also note that in the inertial subrange both $f_L(\kappa L)$ [–] and $f_\eta(\kappa\eta)$ [–] approach near unity so the Kolmogorov $-5/3$ spectrum is produced by the model spectrum. Figure 8.8 shows the energy spectrum function for velocity components U and W [m s^{-1}] from a large-eddy simulation (LES) of anisotropic atmospheric flows [15]. Attempt to identify the slopes for the energy-containing and inertial subranges.

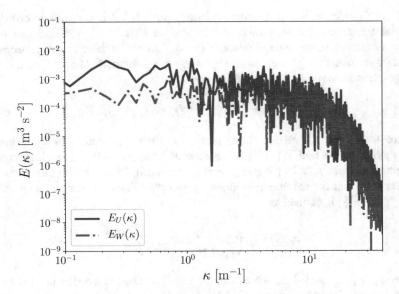

Fig. 8.8 Energy spectrum function from a large-eddy simulation (LES) showing spectrum slopes for the energy-containing and inertial subranges suitable for atmospheric flows: base ten logarithm of energy spectrum function $E(\kappa)$ [m^{3}s^{-2}] versus base ten logarithm of wave number κ [m^{-1}] [15]

References

1. Pope S B (2000) Turbulent flows. Cambridge University Press, Cambridge
2. Kolmogorov A N (1941) The local structure of turbulence in incompressible viscous fluid for very large Reynolds numbers. Dokl Akad Nauk SSSR 30:301–305
3. Aliabadi A A, Moradi M, Byerlay R A E (2021) The budgets of turbulence kinetic energy and heat in the urban roughness sublayer. Environ Fluid Mech 21:843–884
4. Davidson P A (2005) Turbulence: an introduction for scientists and engineers. Oxford University Press, Oxford
5. Aliabadi A A, Staebler R M, Liu M et al (2016) Characterization and parametrization of Reynolds stress and turbulent heat flux in the stably-stratified lower Arctic troposphere using aircraft measurements. Bound-Lay Meteorol 161:99–126
6. Hocking W K (1999) The dynamical parameters of turbulence theory as they apply to middle atmosphere studies. Earth Planets Space 51:525–541
7. Dehghan A, Hocking W K, Srinivasan R (2014) Comparisons between multiple in-situ aircraft turbulence measurements and radar in the troposphere. J Atmos Sol-Terr Phy 118:64–77
8. Hussein H J, Capp S P, George W K (1994) Velocity measurements in a high-Reynolds-number momentum-conserving, axisymmetric, turbulent jet. J Fluid Mech 258:31–75
9. Willis G E, Deardorff J W (1976) On the use of Taylor's translation hypothesis for diffusion in the mixed layer. Q J Roy Meteor Soc 102:817–822
10. Stull R B (1988) An introduction to boundary layer meteorology. Kluwer Academic Publishers, Dordrecht
11. Taylor G I (1938) The spectrum of turbulence. Proc Roy Soc Lond A 164:476–490
12. von Kármán T (1948) Progress in the statistical theory of turbulence. Proc Natl Acad Sci U S A 34:530–539

13. Kaimal J C, Wyngaard J C, Izumi Y et al (1972) Spectral characteristics of surface-layer turbulence. Q J Roy Meteor Soc 98:563–589
14. Kaimal J C, Wyngaard J C, Haugen D A et al (1976) Turbulence structure in the convective boundary layer. J Atmos Sci 33:2152–2169
15. Aliabadi A A, Veriotes N, Pedro G (2018) A Very Large-Eddy Simulation (VLES) model for the investigation of the neutral atmospheric boundary layer. J Wind Eng Ind Aerodyn 183:152–171

Chapter 9
Time and Frequency Domains

Abstract This chapter is concerned with representation of turbulent flow in both time and frequency domains. The discrete Fourier transform is introduced as a technique to transform a turbulence signal, such as a component of the velocity vector field, from the time domain to the frequency domain and vice versa. The Nyquist frequency is defined, which characterizes the highest frequency for fluctuations that is possible to reconstruct from a time series reliably. The discrete energy and energy density spectra are defined with their relationships to the discrete Fourier transform. Finally, the spectra of two variables and the associated mathematical tools to analyse them simultaneously are discussed.

9.1 Overview

Physical phenomena that are represented as a function of time, i.e. in time domain, can also be represented as a function of frequency, i.e. in frequency domain. This is true since most time-evolving phenomena exhibit various repeating patterns that occur at various frequencies. The frequency domain is the description of such repetitions that occur at different levels of strength. The methodology developed in this chapter is adapted from [1].

9.2 Discrete Fourier Transform

From Fourier analysis in calculus we know that any well-behaved continuous function can be described by an infinite Fourier series, i.e. the sum of an infinite number of sine and cosine terms. However, most turbulence signals that are measured and represented with time series are discrete functions of time. In the case of discrete time series with a finite number of points, we only need a finite number of sine and cosine terms to fit our time series points *exactly*, i.e. we can have a perfect transformation of a signal from the *time domain* into the *frequency domain*.

© The Author(s), under exclusive license to Springer Nature Switzerland AG 2022   97
A. A. Aliabadi, *Turbulence*, Mechanical Engineering Series,
https://doi.org/10.1007/978-3-030-95411-6_9

We begin with a time series function $A(k)$ [–], with N [–] discrete timestamps from $k = 0$ to $k = N - 1$ [–] with a discretization of Δt [s]. This time series fills an entire period of time equal to $\mathcal{P} = N \Delta t$ [s]. The idea is to represent this time series function in frequency domain as a sum of sines and cosines with amplitudes $F_A(n)$ [–]. $F_A(n)$ [–] is a complex number, where the real part represents the amplitude of the cosine waves and the imaginary part is the sine waves' amplitude. $F_A(n)$ [–] is a function of frequency n [–] because the waves of different frequencies must be multiplied by different amplitudes to reconstruct the original time series. n [–] is the frequency in the number of cycles per time period $\mathcal{P}$ [s], and $F_A(n)$ [–] is the *discrete Fourier transform* of function $A(k)$ [–] and hence is written with subscript A [–]. There are a number of ways to describe frequency

$$n = \text{number of cycles per entire time period of signal } \mathcal{P}, \tag{9.1}$$

$$f = \text{number of cycles per second} = \frac{n}{\mathcal{P}} = \frac{n}{N \Delta t}, \tag{9.2}$$

$$\omega = \text{radians per second} = 2\pi f = \frac{2\pi n}{N \Delta t}. \tag{9.3}$$

A frequency of zero ($n = 0$) [–] denotes a mean value. The *fundamental frequency*, where $n = 1$ [–], means that exactly one wave fills the entire time period $\mathcal{P}$ [s]. Higher frequencies correspond to *harmonics* of the fundamental frequency. For instance, $n = 5$ [–] means that exactly 5 waves fill the period $\mathcal{P}$ [s].

Using Euler's (1707–1783) formula, $e^{ix} = \cos(x) + i \sin(x)$, as a short notation for sines and cosines, we can write the *Fourier transform pair* to express the relationship between $A(k)$ [–] and $F_A(n)$ [–] using the *discrete Fourier transform*

$$F_A(n) = \sum_{k=0}^{N-1} \left[\frac{A(k)}{N} \right] e^{-i2\pi nk/N}, \tag{9.4}$$

$$A(k) = \sum_{n=0}^{N-1} F_A(n) e^{i2\pi nk/N}, \tag{9.5}$$

which are termed the *forward transform* and *inverse transform*, respectively. If the original time series $A(k)$ [–] is known, then the $F_A(n)$ [–] coefficients can be found from the forward transform, and once these coefficients are known, it is possible to reconstruct the time series using the inverse transform. Conversion of a signal from the time domain to the frequency domain is also known as the *Fourier decomposition*. These expressions state that for a discrete time series with no more than N [–] data points, we do not need more than N [–] different frequencies to describe it. In fact we need less than N [–] frequencies.

Table 9.1 Specific humidity measurements in an atmospheric turbulence study

Index (k)	0	1	2	3	4	5	6	7
Time (UTC)	1200	1215	1230	1245	1300	1315	1330	1345
Q [gkg^{-1}]	8	9	9	6	10	3	5	6

Table 9.2 Coefficients for the forward discrete Fourier transform of specific humidity measurements in an atmospheric turbulence study

n [–]	$F_Q(n)$ [gkg^{-1}]
0	7.0
1	$0.28 - 1.03i$
2	0.5
3	$-0.78 - 0.03i$
4	1.0
5	$-0.78 + 0.03i$
6	0.5
7	$0.28 + 1.03i$

An example can be provided. Eight data points of specific humidity Q [gkg^{-1}] have been collected as a function of time for a turbulence measurement in the atmosphere using a fixed probe. The data points are presented in Table 9.1.

By performing the forward Fourier transform (Eq. 9.4), we can obtain eight coefficients $F_Q(n)$ [gkg^{-1}]. These coefficients are complex, i.e. $F_Q(n) = F_{real}(n) + i F_{imag}(n)$ [gkg^{-1}]. Since $N = 8$ [–] and $\Delta t = 15$ min, the total period is $\mathcal{P} = N\Delta t = 2$ hr. The eight coefficients can be calculated using the following formula and are shown in Table 9.2.

$$F_Q(n) = \frac{1}{N} \sum_{k=0}^{N-1} Q(k) \cos(2\pi nk/N) - \frac{i}{N} \sum_{k=0}^{N-1} Q(k) \sin(2\pi nk/N) . \quad (9.6)$$

This calculation can be checked by the inverse transform (Eq. 9.5) to ensure that the time series signal can be recovered using the following formula:

$$Q(k) = \sum_{n=0}^{N-1} F_{real}(n) \cos(2\pi nk/N) - \sum_{n=0}^{N-1} F_{imag}(n) \sin(2\pi nk/N) . \quad (9.7)$$

These inverse transform sums are in fact four sums as opposed to two sums. The remaining two sums consist of the real part of F times the imaginary factor $i \sin(\ldots)$, and the imaginary part of F [gkg^{-1}] times the real factor $\cos(\ldots)$. Because the last half of the Fourier transforms are the complex conjugates of the first half (not counting the mean), these two sums identically cancel, leaving the two listed above, i.e. Eqs. 9.6 and 9.7. By performing the calculation for the above sum, it can be verified that the signal $Q(k)$ [gkg^{-1}] can be recovered.

9.3   Nyquist Frequency

Almost all measured time series are discrete in nature because every measurement probe has a sampling frequency and cannot measure continuously and infinitely fast. The rule of thumb in discrete data analysis is that at least two data points are needed per period or wavelength in order to resolve a wave with that period or wavelength. Discrete Fourier analysis involves decomposing a time series signal into waves of different frequencies. If we have a total of N [–] data points with a constant sampling frequency, then the highest frequency that we can resolve in our Fourier transform is $n_f = N/2$ [–], also called the *Nyquist frequency*, named after Harry Nyquist (1889– 1976). In other words, if a wave period as small as 0.1 s must be measured, then the time series must be digitized at least every 0.05 s. Similarly, when flying a moving probe into a turbulent flow, if a wavelength as small as 1 m is to be measured, then the signal must be digitized at least once every 0.5 m.

If higher frequencies in a time series or other signals are present but cannot be measured due to the limited sampling frequency of a probe, then those frequencies that are higher than the *Nyquist frequency* are *folded* or *aliased* into lower frequencies. In other words, our analysis will show artificially higher amplitudes for low frequency decomposition of the signal. Common strategies to eliminate folding or aliasing are to use an analog electronic low pass filter to remove high frequencies from the signal before it is sampled. Alternatively, the time series can be block-averaged, for instance by averaging ten adjacent data points.

9.4   Discrete Energy Spectrum

In turbulence studies, it is often desired to know how much of the variance of a fluctuating time series signal is associated with a particular frequency or a range of frequencies. For instance, high-resolution atmospheric boundary layer models are assessed for their ability to resolve flow fluctuations at a particular range of frequencies [2, 3]. The answer to this question is possible using the discrete Fourier transform. The square of the norm of the complex Fourier transform for any frequency n [–] is given by

$$|F_A(n)|^2 = [F_{real}(n)]^2 + [F_{imag}(n)]^2 . \qquad (9.8)$$

where $|F_A(n)|^2$ [–] is summed over frequencies $n = 1$ to $N-1$ [–], the result equals the total biased variance of the original time series, i.e.

$$\sigma_A^2 = \frac{1}{N} \sum_{k=0}^{N-1} (A(k) - \langle A(k) \rangle_T)^2 = \sum_{n=1}^{N-1} |F_A(n)|^2 , \qquad (9.9)$$

where the time average $\langle\rangle_T$ is the only average available to us for calculating the variance. Note that the square of the norm of the complex Fourier transform is summed starting at $n = 1$ instead of $n = 0$ [–]. This is trivial since there are no turbulent fluctuations associated with $n = 0$.

We can interpret $|F_A(n)|^2$ [–] as the portion of variance explained by waves of frequency n [–]. Likewise, the contribution of frequencies from $n = n_1$ to $n = n_2$ [–] to the variance can be found out by summing $|F_A(n)|^2$ [–] from $n = n_1$ to $n = n_2$ [–].

For frequencies greater than the Nyquist frequency, the $|F_A(n)|^2$ [–] values are identically equal to those at the corresponding folded lower frequencies because the Fourier transforms of high frequencies are the same as those for the low frequencies, except for a sign change in the imaginary part. Frequencies higher than the Nyquist frequency cannot be resolved by Fourier transform; therefore, $|F_A(n)|^2$ [–] values at high frequencies should be folded back and added to those at the lower frequencies. Therefore, the *discrete spectral intensity* or *discrete spectral energy*, $E_A(n)$ [–], is defined as $E_A(n) = 2|F_A(n)|^2$ [–], for $n = 1$ to $n = n_f$ [–], with N [–] being odd, while for N [–] being even, $E_A(n) = 2|F_A(n)|^2$ [–] for frequencies from $n = 1$ to $n = n_f - 1$ [–], but $E_A(n) = |F_A(n)|^2$ [–] for $n = n_f$ [–].

9.5   Discrete Energy Density Spectrum

Related to the concept of energy is energy density. If we divide $E_A(n)$ [–] by Δn [–], we obtain the *spectral energy density*

$$S_A(n) = \frac{E_A(n)}{\Delta n}, \tag{9.10}$$

which has the units of A [–] squared per unit frequency. The advantage of spectral energy density is that instead of summing the discrete spectral energy over a range for n [–] to yield variance for that range, one can simply integrate $S_A(n)$ [–] over the same range if a well-defined function can be fitted to $S_A(n)$ [–]. For the entire range of n [–] the variance of A [–] can be given by

$$\sigma_A^2 = \int_1^{n_f} S_A(n)dn . \tag{9.11}$$

For the previous example, having $|F_Q(n)|$ [gkg^{-1}], it is possible to calculate the spectral energy and spectral density according to Table 9.3. Note that in this example N [–] is even so the spectral energy at $n = n_f$ [–] has not been multiplied by two. Also note that $|F_Q(n)|^2$ [g^{2}kg^{-2}] is symmetric about $n = n_f$ [–] showing the nature of high frequencies that have been folded on lower frequencies. It can be verified that the sum of the $|F_Q(n)|^2$ [g^{2}kg^{-2}] or spectral energy is indeed equal to the total biased variance of $Q(k)$ [gkg^{-1}].

Table 9.3 Spectral energy and spectral density for specific humidity measurements in an atmospheric turbulence study

n [–]	$F_Q(n)$ [gkg^{-1}]	$|F_Q(n)|^2$ [g^{2}kg^{-2}]	$E_Q(n)$ [g^{2}kg^{-2}]	$S_Q(n)$ [g^{2}kg^{-2}]
0	7.0 (=mean)			
1	$0.28 - 1.03i$	1.14	2.28	2.28
2	0.5	0.25	0.5	0.5
3	$-0.78 - 0.03i$	0.61	1.22	1.22
$4 = n_f$	1.0	1.0	1.0	1.0
5	$-0.78 + 1.03i$	0.61		
6	0.5	0.25		
7	$0.28 + 1.03i$	1.14		
	Sum	5.0	5.0	5.0

9.6 Spectra of Two Variables

In turbulence studies, such as those involving the atmospheric boundary layer, it is often very useful to analyse the spectrum for a product of two variables [4–8], for instance the product of the fluctuations of velocities in the directions of e_1 and e_2 coordinate vectors, i.e. $u_1 u_2$ [m^{2}s^{-2}]. *Cross-spectrum* analysis relates the spectra of two variables. The *phase* refers to the position within one wave, such as the crest or the trough, and is given as an angle. *Phase shift* refers to the angle between one part of a wave, such as a crest, to that of another wave. The equation for a single sine wave of amplitude C [–] that is shifted by angle $\Phi(n)$ [–] to the right is

$$A(k, n) = C(n) \sin\left(\frac{2\pi kn}{N} - \Phi(n)\right) . \tag{9.12}$$

In this equation the k [–] index signifies time stamp, i.e. represents a particular data point of a time series, while n [–] is a measure of the frequency of the wave. The same equation can be shown as the sum of the sine and cosine waves by

$$A(k, n) = C_s(n) \sin\left(\frac{2\pi kn}{N}\right) + C_c(n) \cos\left(\frac{2\pi kn}{N}\right) , \tag{9.13}$$

where $C_s(n) = C(n) \cos(\Phi(n))$ and $C_c(n) = -C(n) \sin(\Phi(n))$ [–]. As was discussed, the Fourier transforms give the amplitudes of sine and cosine terms in the spectral decomposition of the original time series. Therefore, we can also interpret the spectra in terms of an amplitude and phase shift for waves of each frequency.

We can define $G_A(n) = |F_A(n)|^2$ or $G_A = |F_A|^2$ [–] for short as the unfolded spectral energy for variable A [–] and frequency n [–]. We can write $G_A = F_A^* \cdot F_A$ [–], where F_A^* [–] is the complex conjugate of F_A [–], and where the dependence on n [–] is still implied. This can be shown by considering the real and imaginary parts of F_A [–], i.e. $F_A = F_{Ar} + i F_{Ai}$ [–], where subscripts r and i signify the real and imaginary parts, respectively,

$$
\begin{aligned}
G_A &= F_A^* . F_A \\
&= (F_{Ar} - i F_{Ai})(F_{Ar} + i F_{Ai}) \\
&= F_{Ar}^2 + i F_{Ai} F_{Ar} - i F_{Ai} F_{Ar} - i^2 F_{Ai}^2 \\
&= F_{Ar}^2 + F_{Ai}^2 \\
&= |F_A|^2 .
\end{aligned}
\tag{9.14}
$$

We can now define the *cross-spectrum* between two variables A [–] and B [–] by

$$
\begin{aligned}
G_{AB} &= F_A^* . F_B \\
&= (F_{Ar} - i F_{Ai})(F_{Br} + i F_{Bi}) \\
&= F_{Ar} F_{Br} + i F_{Ar} F_{Bi} - i F_{Ai} F_{Br} - i^2 F_{Ai} F_{Bi} .
\end{aligned}
\tag{9.15}
$$

If we now collect the real parts and the imaginary parts, the real part is defined as the *cospectrum*, Co [–], and the imaginary part is defined as *quadrature spectrum*, Q [–]

$$
G_{AB} = Co + i Q ,
\tag{9.16}
$$

where

$$
Co = F_{Ar} F_{Br} + F_{Ai} F_{Bi} ,
\tag{9.17}
$$

$$
Q = F_{Ar} F_{Bi} - F_{Ai} F_{Br} .
\tag{9.18}
$$

It must be reminded that both F_A [–] and F_B [–] are still functions of frequency n [–], resulting in both cospectrum and quadrature spectrum to be also functions of frequency n [–], i.e. $Co(n)$ and $Q(n)$ [–].

The cospectrum is frequently used in turbulence studies because the sum over all frequencies of cospectrum amplitudes is equal to the covariance of turbulent fluctuations in A and B [–], i.e. a and b [–], such that

$$
\sum_n Co(n) = \langle ab \rangle_T ,
\tag{9.19}
$$

where again it is assumed that only time averaging is possible for time series as a close approximate for covariance. It is important to note that the cospectrum computed this way is not equal to the spectrum of the time series of the product ab [–].

Problems

9.1 A time series has a period of $\mathcal{P} = 15\,\text{s}$. This time series is composed of numerous harmonic frequencies. Calculate the third harmonic frequency, i.e. $n = 3$ [–], in units of cycles per second (f [s^{-1}]) and radians per second (ω [Rads^{-1}]).

9.2 A time series signal has $N = 100$ [–] timestamps with a constant sampling interval of $\Delta t = 2\,\text{s}$. Show that the highest frequency that a Fourier transform of this signal can resolve, i.e. the Nyquist frequency, is $n_f = 50$ [–] cycles per the entire time period of the signal, or equivalently $f = 0.25\,\text{Hz}$ cycles per second.

9.3 What is folding or aliasing?

9.4 The spectral energy density for a component of momentum, in [m^{2}s^{-2} per the number of cycles per the entire time period of the flow], for a turbulent flow in part of the dissipation subrange is given by the following expression:

$$S(n) = \frac{100}{1 + n^2} , \tag{9.20}$$

where n [–] is the number of cycles per entire time period of the flow. Note that this expression is only valid from $n = 100$ to $n = 200$ [–]. Calculate the amount of variance σ^2 for the component of momentum in this flow associated with the number of cycles per entire time period of the flow from $n_1 = 120$ to $n_2 = 140$ [–]. What is the unit for the calculated variance?

References

1. Stull R B (1988) An introduction to boundary layer meteorology. Kluwer Academic Publishers, Dordrecht
2. Aliabadi A A, Veriotes N, Pedro G (2018) A Very Large-Eddy Simulation (VLES) model for the investigation of the neutral atmospheric boundary layer. J Wind Eng Ind Aerodyn 183:152–171
3. Ahmadi-Baloutaki M, Aliabadi A A (2021) A very large-eddy simulation model using a reductionist inlet turbulence generator and wall modeling for stable atmospheric boundary layers. Fluid Dyn 56:413–432
4. Kaimal J C, Finnigan J J (1994) Atmospheric boundary layer flows: their structure and measurement. Oxford University Press, Oxford
5. Kaimal J C, Wyngaard J C, Izumi Y et al. (1972) Spectral characteristics of surface-layer turbulence. Q J Roy Meteor Soc 98:563–589
6. Kaimal J C, Wyngaard J C, Haugen D A et al (1976) Turbulence structure in the convective boundary layer. J Atmos Sci 33:2152–2169
7. Aliabadi A A, Moradi M, Clement D et al (2019) Flow and temperature dynamics in an urban canyon under a comprehensive set of wind directions, wind speeds, and thermal stability conditions. Environ Fluid Mech 19:81–109
8. Aliabadi A A, Moradi M, Byerlay R A E (2021) The budgets of turbulence kinetic energy and heat in the urban roughness sublayer. Environ Fluid Mech 21:843–884

Part II
Measurement Techniques

Chapter 10
Fundamentals of Measurements

Abstract This chapter introduces the fundamentals of measurements in experiments. Significant digits are introduced, which are digits of a number that can be measured reliably in an experiment. The concept of calibration is discussed, which involves validating a set of measurements against a set of reference measurements. Precision and accuracy are defined as different measures of uncertainty. Mean and standard deviation are discussed as statistical tools used to report measurement uncertainties. Normal and student's t probability density functions are defined, which are used in the expression of uncertainty and calculation of confidence levels in measurements. A technique in rejection of unreliable data is provided. Least-square fitting is introduced. The Chi-squared test for a distribution is provided. The two sample statistical estimation test is provided to compare two sets of measurements characterized by their mean and standard deviation. Finally, uncertainty reporting and the propagation of uncertainty are discussed.

10.1 Overview

Before introducing individual thermo-fluids measurement techniques, it is essential to introduce fundamentals of measurements. Example topics to include are significant digits, calibration, uncertainty, statistical analysis of random uncertainties, normal and Student's t probability distribution density functions, rejection of data, least-square fitting, Chi-squared test for a distribution, two sample statistical estimation, reporting of uncertainties, and propagation of uncertainties. The material in this chapter is adapted from standard texts in probability and statistics [1, 2].

10.2 Significant Digits

In engineering and science, significant digits refer to digits of a number that can be reliably measured. In another word, the number of significant digits refers to the uncertainty of a measurement. For instance, if the uncertainty of a measurement is

±0.04, then the measurement should only contain as many decimal places as the uncertainty, e.g. 2.31 not 2.3124.

Some rules govern the quantification of the number of significant digits. All non-zero digits are considered significant. For instance the number 564.32 has 5 significant digits. Zeros appearing between two non-zero digits are significant. For instance the number 1.0003 has 5 significant digits. Leading zeros are not significant. For instance, 0.00013 has two significant digits. If included, trailing zeros in a number are significant digits. For instance 12,300 has 5 significant digits. However, if trailing zeros are not significant, then scientific notation must be used, so that trailing zeros are not interpreted as significant digits. For the same example, then 1.23×10^4 contains only 3 significant digits.

When multiplying or dividing two numbers, the result should have as many significant digits as the measured number with the smallest number of significant digits. For instance 2.5×3.42 should be written as 8.6 not 8.55. Also when adding or subtracting two numbers, the result should have as many significant digits as the measured number with the smallest number of significant digits. For instance 1.23 + 4.567 should be written as 5.80 not 5.797.

The scientific notation is used to remove the non-significant digits from a number. Typically, only 1 digit to the left of the decimal point is used. For instance 0.00012 should be reported using 1.2×10^{-4} not 12×10^{-5}.

To remove non-significant digits from a calculated number, rounding can be used. In this technique the last number is eliminated, and the one before the last number is incremented by 1 if the last number is 5 and above and is kept the same if the last number is 4 or less.

10.3   Calibration

Errors arise in the measured values because the instruments and measurement procedure are not perfect, i.e. output of the measurement does not precisely follow the standard measurement. These errors can be quantified through the process of calibration. In calibration, a known standard is applied to the system and the system output is corrected given the standard measurement.

Figure 10.1 shows a simple calibration curve. In this exercise a set of measured values M [–] are compared to a set of corresponding standard values S [–]. Then a calibration curve is fitted, so that it takes as input the M [–] values and it produces the expected S [–] values. In this example the calibration curve is a line $S = aM + b$ [–], but non-linear calibration curves may also be used (e.g. polynomial, exponential).

Fig. 10.1 Calibration curve
of measured versus standard
values

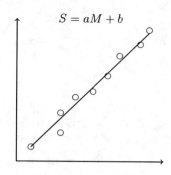

Standard values S

$$S = aM + b$$

Measured values M

10.4   Uncertainty

The uncertainty is an estimate of the effects of errors on the results of the measurement. Uncertainty includes all of the errors that are present in the measurement
system, calibration, and measurement technique. Uncertainty can include either
random errors or systematic errors. Random errors are caused by the lack of
repeatability in the measurement system. The scatter in data represents the random
error. Random errors may be caused by the experimental system, instruments, or the
environment. By analysing the spread in results statistically, we can compute a very
reliable estimate of this kind of error. Systematic errors are often not obvious to the
experimenter, and they cannot be discovered by statistical analysis. Only calibration
can identify systematic errors.

Random and systematic errors are also quantified using precision and accuracy.
A high precision measurement exhibits a low random error, and a high accuracy measurement exhibits a low systematic error. Figure 10.2 shows the visual
representation of random and systematic errors. In this figure the measurement
attempts to be on target, i.e. the centre circle. A measurement with low random
error (precise measurement) results in a cluster of points in the same vicinity, while a
measurement with low systematic error (accurate measurement) results in the cluster
of measurements to be on target on average. This figure shows four possibilities: (1)
precise and accurate, (2) precise but not accurate, (3) not precise but accurate, and
(4) neither precise nor accurate.

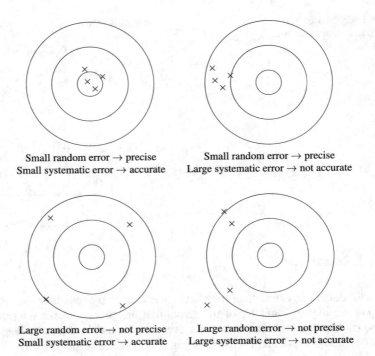

Fig. 10.2 Visual representation of random errors (precision) and systematic errors (accuracy)

10.5 Statistical Analysis of Random Uncertainties

Real measurements usually present a degree of randomness caused by variations in uncontrolled variables. In fact a high level of randomness may completely obscure a trend or relationship in the data. Statistical analysis of data enables conclusions to be drawn about the data trends. In this approach, multiple measurements of a physical variable are treated as a random variable. Although the statistical mean of a random variable may never be known, a good estimator for the mean is the average, defined over N [–] measurements of random variable x [–] as

$$\overline{x} = \frac{x_1 + x_2 + \cdots + x_N}{N}. \qquad (10.1)$$

The standard deviation or root mean square deviation of the measurements is an estimate of the average random uncertainty of the measurement. Here the deviation is taken with respect to the best available estimator of the mean, i.e. the average.

Over large enough data sets, i.e. if $N > 20$ [–], the standard deviation is defined as

$$\sigma_x = \sqrt{\frac{\sum_{i=1}^{N} (x_i - \overline{x})^2}{N}}. \tag{10.2}$$

For smaller data sets, i.e. $N < 20$ [–], a more conservative relation for standard deviation is used as the following, also known as the sample standard deviation,

$$\sigma_x = \sqrt{\frac{\sum_{i=1}^{N} (x_i - \overline{x})^2}{N - 1}}. \tag{10.3}$$

The variance is simply the square of the standard deviation. The median value is another estimator that provides the central tendency in a number of measurements. It is the value at the centre of a set of measurements. That is, half of the values fall above and the other half below this central value. The mode is the most frequently occurring event, so that the probability of measuring the mode in a system is the highest probability.

The standard deviation σ_x [–] represents the average random uncertainty of separate measurements. If multiple averages are taken, say $\overline{x}_1, \overline{x}_2, \ldots$ [–], each over N [–] number of measurements, then the standard deviation of the random variable $\overline{x}$ [–] (with the best estimator denoted as $\overline{\overline{x}}$ [–]) will be lower than the standard deviation of the random variable x [–] (with the best estimator denoted as $\overline{x}$ [–]) such that

$$\sigma_{\overline{x}} = \frac{\sigma_x}{\sqrt{N}}, \tag{10.4}$$

which is also known as the standard deviation of the average.

10.6 Normal and Student's *t* Distributions

Numerous measurements can be represented using probability distribution density functions. Many measurements are found to have a symmetric bell-shaped curve for their limiting distribution, i.e. the case where $N \geq 30$ [–], also known as the normal distribution or Gaussian distribution after the German mathematician Johann Carl Friedrich Gauss (1777–1855). In these measurements the errors are random, so random uncertainty is best represented by this kind of distribution. Note that in this scenario the number of observations above and below the average value will be the same. Even though a systematic uncertainty may exist, it pushes all measured values in one direction, so the shape of the distribution is not affected by the presence of the systematic uncertainty. Therefore, the distribution will be off-centred with respect to the true mean. If the true mean μ [–] (not average $\overline{x}$ [–] in Eq. 10.1) and width

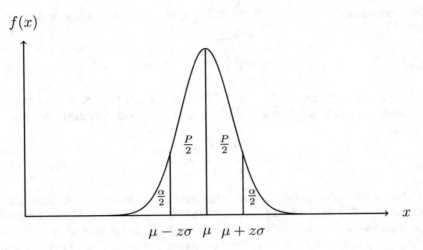

Fig. 10.3 The normal distribution probability density function $f(x)$ with mean μ and width σ

parameter σ [–] (not standard deviation σ_x [–] in Eq. 10.2) are known, the normal or Gaussian distribution is given by the following expression:

$$f(x) = \frac{1}{\sigma\sqrt{2\pi}} \exp\left[-\frac{(x-\mu)^2}{2\sigma^2}\right]. \tag{10.5}$$

The fundamental property of probability distribution density functions requires that $\int_{-\infty}^{\infty} f(x)dx = 1$ [–]. However, here our interest is to state, with some level of confidence P [–], in which range a measured random variable will be. With the aid of Fig. 10.3, we can note that the confidence level P [–] corresponds to an area $\frac{P}{2}$ [–] to the right of μ to $\mu + z\sigma$ [–], and it corresponds to an area $\frac{P}{2}$ [–] to the left of μ to $\mu - z\sigma$ [–]. So with this confidence level the measured value will be between $\mu - z\sigma$ and $\mu + z\sigma$ [–]. Further, the total area satisfies $\frac{P}{2} + \frac{P}{2} + \frac{\alpha}{2} + \frac{\alpha}{2} = 1$ [–]; in other words $P = 1 - \alpha$ and $\alpha = 1 - P$ [–]. In fact $\frac{P}{2}$ [–] is a function of z [–] and vice versa, so that if a value of $\frac{P}{2}$ [–] is desired (e.g. $\frac{P}{2} = 0.45$ [–] for 90% confidence level), then a value of z [–] can be looked up from a table (for $\frac{P}{2} = 0.45$ [–], we can look up $z \approx 1.64$ [–]). Table 10.1 is known as the z table and provides the relationship between $\frac{P}{2}$ [–] and z [–] for a normal distribution with true mean μ [–] and width σ [–]. The first column of this table shows the first two significant digits of z [–], while the first row shows the second decimal place of z [–]. The numbers in other columns and rows are values of $\frac{P}{2}$ [–]. For instance if $\frac{P}{2} = 0.45$ [–] is desired, the closest numbers in Table 10.1 would be $\frac{P}{2} = 0.4495, 0.4505$ [–] corresponding to $z = 1.64, 1.65$ [–], respectively.

Note that in a finite sample of measurements, the true mean μ [–] and width σ [–] of the normal distribution are not available. In this case the best available estimators are the average $\overline{x}$ [–] (for μ [–]), given by Eq. 10.1, and standard deviation σ_x [–]

Table 10.1 The percentage probability $\frac{P}{2}$ [–] that the random variable x [–] lies between the true mean μ and $\mu + z\sigma$ [–] in a normal distribution with true mean μ and width σ [–]

z	0.00	0.01	0.02	0.03	0.04	0.05	0.06	0.07	0.08	0.09
0.0	0.0000	0.0040	0.0080	0.0120	0.0160	0.0199	0.0239	0.0279	0.0319	0.0359
0.1	0.0398	0.0438	0.0478	0.0517	0.0557	0.0596	0.0636	0.0675	0.0714	0.0753
0.2	0.0793	0.0832	0.0871	0.0910	0.0948	0.0987	0.1026	0.1064	0.1103	0.1141
0.3	0.1179	0.1217	0.1255	0.1293	0.1331	0.1368	0.1406	0.1443	0.1480	0.1517
0.4	0.1554	0.1591	0.1628	0.1664	0.1700	0.1736	0.1772	0.1808	0.1844	0.1879
0.5	0.1915	0.1950	0.1985	0.2019	0.2054	0.2088	0.2123	0.2157	0.2190	0.2224
0.6	0.2257	0.2291	0.2324	0.2357	0.2389	0.2422	0.2454	0.2486	0.2517	0.2549
0.7	0.2580	0.2611	0.2642	0.2673	0.2704	0.2734	0.2764	0.2794	0.2823	0.2852
0.8	0.2881	0.2910	0.2939	0.2967	0.2995	0.3023	0.3051	0.3078	0.3106	0.3133
0.9	0.3159	0.3186	0.3212	0.3238	0.3264	0.3289	0.3315	0.3340	0.3365	0.3389
1.0	0.3413	0.3438	0.3461	0.3485	0.3508	0.3531	0.3554	0.3577	0.3599	0.3621
1.1	0.3643	0.3665	0.3686	0.3708	0.3729	0.3749	0.3770	0.3790	0.3810	0.3830
1.2	0.3849	0.3869	0.3888	0.3907	0.3925	0.3944	0.3962	0.3980	0.3997	0.4015
1.3	0.4032	0.4049	0.4066	0.4082	0.4099	0.4115	0.4131	0.4147	0.4162	0.4177
1.4	0.4192	0.4207	0.4222	0.4236	0.4251	0.4265	0.4279	0.4292	0.4306	0.4319
1.5	0.4332	0.4345	0.4357	0.4370	0.4382	0.4394	0.4406	0.4418	0.4429	0.4441
1.6	0.4452	0.4463	0.4474	0.4484	0.4495	0.4505	0.4515	0.4525	0.4535	0.4545
1.7	0.4554	0.4564	0.4573	0.4582	0.4591	0.4599	0.4608	0.4616	0.4625	0.4633
1.8	0.4641	0.4649	0.4656	0.4664	0.4671	0.4678	0.4686	0.4693	0.4699	0.4706
1.9	0.4713	0.4719	0.4726	0.4732	0.4738	0.4744	0.4750	0.4756	0.4761	0.4767
2.0	0.4772	0.4778	0.4783	0.4788	0.4793	0.4798	0.4803	0.4808	0.4812	0.4817
2.1	0.4821	0.4826	0.4830	0.4834	0.4838	0.4842	0.4846	0.4850	0.4854	0.4857
2.2	0.4861	0.4864	0.4868	0.4871	0.4875	0.4878	0.4881	0.4884	0.4887	0.4890
2.3	0.4893	0.4896	0.4898	0.4901	0.4904	0.4906	0.4909	0.4911	0.4913	0.4916
2.4	0.4918	0.4920	0.4922	0.4925	0.4927	0.4929	0.4931	0.4932	0.4934	0.4936
2.5	0.4938	0.4940	0.4941	0.4943	0.4945	0.4946	0.4948	0.4949	0.4951	0.4952
2.6	0.4953	0.4955	0.4956	0.4957	0.4959	0.4960	0.4961	0.4962	0.4963	0.4964
2.7	0.4965	0.4966	0.4967	0.4968	0.4969	0.4970	0.4971	0.4972	0.4973	0.4974
2.8	0.4974	0.4975	0.4976	0.4977	0.4977	0.4978	0.4979	0.4979	0.4980	0.4981
2.9	0.4981	0.4982	0.4982	0.4983	0.4984	0.4984	0.4985	0.4985	0.4986	0.4986
3.0	0.4987	0.4987	0.4987	0.4988	0.4988	0.4989	0.4989	0.4989	0.4990	0.4990

(for σ [–]), given by Eq. 10.2. These can be used instead to provide the following range with confidence level P [–] (and corresponding z [–] for $\frac{P}{2}$ [–]) for the random measurement

$$x = \bar{x} \pm z\sigma_x. \tag{10.6}$$

For small sample sizes, i.e. $N < 30$ [–], statistical values of average and standard deviation obtained from data sets should be regarded as estimates of the true mean and width statistics. In this case the normal distribution is replaced by the

Student's t distribution

$$x = \overline{x} \pm t\sigma_x, \tag{10.7}$$

where the value of t [–] corresponds to $\frac{P}{2}$ [–] and degrees of freedom $\nu = N - 1$ [–] in Table 10.2. In this table, the first column presents the degrees of freedom ν [–], the first row presents $\frac{P}{2}$ [–], and the other rows and columns provide the t-values associated with $\frac{P}{2}$ [–] and ν [–]. It must be noted that if $N \geq 30$ [–] the values of t [–] and z [–] become close to one another for a given $\frac{P}{2}$ [–]. The t distribution is also used for hypothesis testing in statistics [3].

10.7   Rejection of Data

Sometimes, some data points in a series of measurements appear to disagree strikingly in comparison to the bulk of the data. We need to consider if these data points are a result of some gross experimental error, or they represent some new type of physical phenomenon that is peculiar to certain operating condition. There is a simple test called Chauvenet's criterion that can be considered by the experimenter. For N [–] measurements of a single quantity x [–], i.e. $x_1, x_2, \ldots, x_N$ [–], we should calculate $\overline{x}$ [–] and σ_x [–]. For the suspected measurement x_{sus} [–], we then find the number of standard deviations by which x_{sus} [–] differs from $\overline{x}$ [–]

$$z_{sus} = \frac{|x_{sus} - \overline{x}|}{\sigma_x} \text{ or } t_{sus} = \frac{|x_{sus} - \overline{x}|}{\sigma_x}. \tag{10.8}$$

Then we use the normal distribution (z_{sus} [–] for $N \geq 30$ [–]) or Student's t distribution (t_{sus} [–] for $N < 30$ [–]) to find the probability for the number to be outside $\overline{x} \pm z_{sus}\sigma_x$ [–] or $\overline{x} \pm t_{sus}\sigma_x$ [–], for normal or Student's t distributions, respectively. This calculation will be performed for the confidence level P [–], i.e. corresponding $\frac{P}{2}$ [–] for z_{sus} [–] or t_{sus} [–], such that

$$\alpha = 1 - P. \tag{10.9}$$

We should then multiply the above probability by N [–] to obtain $n = \alpha N$ [–]. If n [–] is less than 0.5 [–], then according to Chauvenet's criterion we can reject x_{sus} and recalculate the mean and standard deviation of the sample with the remainder of the data.

Table 10.2 The percentage probability $\frac{P}{2}$ [–] that the random variable x [–] lies between the true mean μ and $\mu + t\sigma$ [–] in a Student's t distribution with true mean μ [–], width σ [–], and degrees of freedom $\nu = N - 1$ [–]

ν	0.25	0.4	0.45	0.475	0.49	0.495	0.4975	0.499	0.4995	0.49975	0.4999	0.49995
1	1.00	3.08	6.31	12.71	31.82	63.66	127.32	318.31	636.62	1273.24	3183.10	6366.20
2	0.82	1.89	2.92	4.30	6.96	9.22	14.09	22.33	31.60	44.70	70.70	99.99
3	0.76	1.64	2.35	3.18	4.54	5.84	7.45	10.21	12.92	16.33	22.20	28.00
4	0.74	1.53	2.13	2.78	3.75	4.60	5.60	7.17	8.61	10.31	13.03	15.54
5	0.73	1.48	2.02	2.57	3.37	4.03	4.77	5.89	6.87	7.98	9.68	11.18
6	0.72	1.44	1.94	2.45	3.14	3.71	4.32	5.21	5.96	6.79	8.02	9.08
7	0.71	1.42	1.90	2.37	3.00	3.50	4.03	4.79	5.41	6.08	7.06	7.88
8	0.71	1.40	1.86	2.31	2.90	3.36	3.83	4.50	5.04	5.62	6.44	7.12
9	0.70	1.38	1.83	2.26	2.82	3.25	3.69	4.30	4.78	5.29	6.01	6.59
10	0.70	1.37	1.81	2.23	2.76	3.17	3.58	4.14	4.59	5.05	5.69	6.21
11	0.70	1.36	1.80	2.20	2.72	3.11	3.50	4.03	4.44	4.86	5.45	5.92
12	0.70	1.36	1.78	2.18	2.68	3.06	3.43	3.93	4.32	4.72	5.26	5.69
13	0.69	1.35	1.77	2.16	2.65	3.01	3.37	3.85	4.22	4.60	5.11	5.51
14	0.69	1.35	1.76	2.15	2.63	2.98	3.33	3.79	4.14	4.50	4.99	5.36
15	0.69	1.34	1.75	2.13	2.60	2.95	3.29	3.73	4.07	4.42	4.88	5.24
16	0.69	1.34	1.75	2.12	2.58	2.92	3.25	3.69	4.02	4.35	4.79	5.13
17	0.69	1.33	1.74	2.11	2.57	2.90	3.22	3.65	3.97	4.29	4.71	5.04
18	0.69	1.33	1.73	2.10	2.55	2.88	3.20	3.61	3.92	4.23	4.65	4.97
19	0.69	1.33	1.73	2.09	2.54	2.86	3.17	3.58	3.88	4.19	4.59	4.90
20	0.69	1.33	1.73	2.09	2.53	2.85	3.15	3.55	3.85	4.15	4.54	4.84

(continued)

Table 10.2 (continued)

ν	0.25	0.4	0.45	0.475	0.49	0.495	0.4975	0.499	0.4995	0.49975	0.4999	0.49995
21	0.69	1.32	1.72	2.08	2.52	2.83	3.14	3.53	3.82	4.11	4.49	4.78
22	0.69	1.32	1.72	2.07	2.51	2.82	3.12	3.51	3.79	4.08	4.45	4.74
23	0.68	1.32	1.71	2.07	2.50	2.81	3.10	3.49	3.77	4.05	4.42	4.69
24	0.68	1.32	1.71	2.06	2.49	2.80	3.09	3.47	3.75	4.02	4.38	4.65
25	0.68	1.32	1.71	2.06	2.49	2.79	3.08	3.45	3.73	4.00	4.35	4.62
26	0.68	1.32	1.71	2.06	2.48	2.78	3.07	3.44	3.71	3.97	4.32	4.59
27	0.68	1.31	1.70	2.05	2.47	2.77	3.06	3.42	3.69	3.95	4.30	4.56
28	0.68	1.31	1.70	2.05	2.47	2.76	3.05	3.41	3.67	3.94	4.28	4.53
29	0.68	1.31	1.70	2.05	2.46	2.76	3.04	3.40	3.66	3.92	4.25	4.51
30	0.68	1.31	1.70	2.04	2.46	2.75	3.03	3.39	3.65	3.90	4.23	4.48

10.8 Least-Squares Fitting

Suppose we have a set of measurements $(x_1, y_1), \ldots, (x_n, y_n)$ [(-,-)]. We wish to find the best function $y_c = h(x)$ [–] to fit this set of measurements. We focus our attention on $h(x)$ [–] to be a polynomial function of mth order, where $m \leq n - 1$ [–] such that

$$y_c = a_0 + a_1 x + a_2 x^2 + \cdots + a_m x^m. \tag{10.10}$$

We assume that the distribution of measured values y_i [–] about each fixed value of the independent variable x_i [–] follows a normal distribution with the same width parameter σ_y [–] for all measurements. Using the polynomial fit, the best estimate of y_i [–] at each x_i [–] will be given by

$$y_{c,i} = a_0 + a_1 x_i + a_2 x_i^2 + \cdots + a_m x_i^m. \tag{10.11}$$

Suppose that $y_{c,i}$ [–] is equal to the mean μ [–] of normal distribution at each x_i [–]. With the assumption of the normal distribution (Eq. 10.5), the probability of obtaining the observed value y_i [–] is

$$f(y_i) = \frac{1}{\sigma_y \sqrt{2\pi}} \exp\left[-\frac{(y_i - y_{c,i})^2}{2\sigma_y^2}\right]. \tag{10.12}$$

The probability of obtaining our complete set of measurements $(y_1, \ldots, y_n)$ [(-,...,-)] is the product $f(y_1, \ldots, y_n) = f(y_1) f(y_2) \ldots f(y_n)$ [–], and, using the normal distribution, will be

$$f(y_1, \ldots, y_n) = \frac{1}{\sigma_y \sqrt{2\pi}} \exp\left[-\frac{\sum_{i=1}^{n}(y_i - y_{c,i})^2}{2\sigma_y^2}\right]. \tag{10.13}$$

The best estimates for coefficients $a_0, a_1, \ldots, a_m$ [–] are those for which the probability $f(y_1, \ldots, y_n)$ [–] is maximized. This is equivalent to minimizing the exponent of the exponential function, i.e.

$$D = \sum_{i=1}^{n}(y_i - y_{c,i})^2. \tag{10.14}$$

In other words, it is desired to find the least squares. This explains why least-squares fitting is used in the first place since it is related to the normal distribution. To find the minimum of $D(a_0, \ldots, a_m)$ [–], we can assume that the multi-variable function $D(a_0, \ldots, a_m)$ [–] is smooth and differentiable. Therefore, its total variation with respect to all of its variables must be zero where it is minimized. This requires each of the partial derivatives to be zero too,

$$dD = \frac{\partial D}{\partial a_0} da_0 + \frac{\partial D}{\partial a_1} da_1 + \cdots + \frac{\partial D}{\partial a_m} da_m = 0, \tag{10.15}$$

$$\frac{\partial D}{\partial a_0} = \frac{\partial}{\partial a_0} \left[\sum_{i=1}^{n} \left(y_i - a_0 - a_1 x_1 - \cdots - a_m x_i^m \right)^2 \right],$$

$$\vdots$$

$$\frac{\partial D}{\partial a_m} = \frac{\partial}{\partial a_m} \left[\sum_{i=1}^{n} \left(y_i - a_0 - a_1 x_1 - \cdots - a_m x_i^m \right)^2 \right]. \tag{10.16}$$

This allows for $m + 1$ [–] coefficients a_m [–] to be calculated using the above equations. For linear polynomials, i.e. line fits $y_c = a_0 + a_1 x$ [–], the coefficients can be calculated from

$$a_0 = \frac{\sum x_i \sum (x_i y_i) - \sum x_i^2 \sum y_i}{\left(\sum x_i \right)^2 - n \sum x_i^2}, \tag{10.17}$$

$$a_1 = \frac{\sum x_i \sum y_i - n \sum x_i y_i}{\left(\sum x_i \right)^2 - n \sum x_i^2}. \tag{10.18}$$

The uncertainty in curve fitting can be obtained using the standard error of the fit. Assuming negligible uncertainty in x [–] measurements, the standard error of the fit can be given by

$$S_{yx} = \sqrt{\frac{\sum_{i=1}^{n} \left(y_i - y_{c,i} \right)^2}{\nu}}, \tag{10.19}$$

where $\nu = n - (m+1)$ [–] is the degrees of freedom of the fit. For negligible random error in the independent variable x [–], a confidence interval (with confidence level P [–]) of the curve fit y_c [–] due to random scatter about the fit can also be estimated using

$$y_c \pm z \frac{S_{yx}}{\sqrt{n}} \text{ or } y_c \pm t \frac{S_{yx}}{\sqrt{n}}, \tag{10.20}$$

where either the normal or Student's t distribution can be used for identifying z or t [–], respectively. If a Student's t distribution is used, the degree of freedom is $\nu = n - (m + 1)$ [–]. For linear polynomials, i.e. line fits, a correlation coefficient r

[–] can be found using

$$r = \sqrt{1 - \frac{S_{yx}^2}{S_y^2}},$$

(10.21)

$$S_y^2 = \frac{1}{n-1} \sum_{i=1}^{n} (y_i - \overline{y})^2,$$

(10.22)

$$\overline{y} = \frac{1}{n} \sum_{i=1}^{n} y_i.$$

(10.23)

Alternatively, the value of r^2 [–] is often reported, which is indicative of how well the variance in y [–] is accounted for by the line fit. It is desirable to have r^2 [–] as close to 1 as possible.

10.9   Chi-Squared Test for a Distribution

The Chi-squared test is used to decide if an observed distribution is consistent with an expected theoretical distribution. If we make N [–] measurements for which we know, or we can calculate, the expected values according to a distribution (e.g. normal, polynomial), then we can compare how well the observed values agree with the expected distribution. The Chi-squared can be calculated for the comparison as

$$\chi^2 = \sum_{i=1}^{n} \left[\frac{\text{Observed values} - \text{Expected values}}{\text{Standard deviation}} \right]^2,$$

(10.24)

where the standard deviation comes from the observations and $n \neq N$ [–] is the number of bins for the comparison. This number is a reasonable indicator of the agreement. If $\chi^2 = 0$ [–], then the expected distribution and observations match perfectly. The larger the χ^2 [–] the smaller the probability that the observed values match the expected distribution. To perform this test, the observed values are divided into n [–] bins. Each bin should contain at least one observation. We calculate the width of each bin using $\Delta x = x_{range}/n$ [–]. We compare Δx [–] with σ_x [–] (calculated based on the N [–] measurements). The number of measurements falling into each bin k [–] is counted and denoted by O_k [–] (observed number). The N [–] measurements here are the sum of numbers $O_1, \ldots, O_n$ [–] (i.e. $N = \sum_{k=1}^{n} O_k$ [–]). The expected number E_k [–] is determined by the assumed distribution of x [–]. For instance if a normal distribution is assumed, its probability density function can be used to calculate the expected number E_k [–] by integrating this density function over each bin k [–]. The standard deviation is in the order of $\sqrt{E_k}$ [–]; therefore, the

Chi-squared can be computed by

$$\chi^2 = \sum_{k=1}^{n} \frac{(O_k - E_k)^2}{E_k}. \tag{10.25}$$

If the assumed distribution of x [–] is correct, then χ^2 [–] should be in the order of n [–] or less. If $\chi^2 \gg n$ [–], then the assumed distribution is probably incorrect. To find the degree of agreement between the observed values and expected distribution, we first calculate the number of degrees of freedom as

$$\nu = n - c, \tag{10.26}$$

where n [–] is the number of bins and c [–] is the number of constraints, i.e. mean, width of the distribution, and expected value. For this type of problem $c = 3$ [–], so $\nu = n - 3$ [–]. Once χ^2 [–] and ν [–] are determined, we then use Table 10.3 to find the level of confidence P [–]. In this table the first column is degrees of freedom ν

Table 10.3 The percentage probability P [–] that the observed values agree with the expected assumed distribution according to the value of χ^2 [–] with degrees of freedom ν [–]

ν	0.995	0.990	0.975	0.950	0.900	0.750	0.500	0.250	0.100	0.050
1	0.00	0.00	0.00	0.00	0.01	0.10	0.45	1.32	2.71	3.84
2	0.01	0.02	0.05	0.10	0.21	0.57	1.39	2.77	4.61	5.99
3	0.07	0.11	0.21	0.35	0.58	1.21	2.37	4.11	6.25	7.81
4	0.20	0.29	0.48	0.71	1.06	1.92	3.36	5.39	7.78	9.49
5	0.41	0.55	0.83	1.15	1.61	2.67	4.35	6.63	9.24	11.1
6	0.67	0.87	1.24	1.64	2.20	3.45	5.35	7.84	10.6	12.6
7	0.98	1.24	1.69	2.17	2.83	4.25	6.35	9.04	12.0	14.1
8	1.34	1.65	2.18	2.73	3.49	5.07	7.34	10.2	13.4	15.5
9	1.73	2.09	2.70	3.33	4.17	5.90	8.34	11.4	14.7	16.9
10	2.16	2.56	3.25	3.94	4.87	6.74	9.34	12.5	16.0	18.3
11	2.60	3.05	3.82	4.57	5.58	7.58	10.3	13.7	17.3	19.7
12	3.07	3.57	4.40	5.23	6.30	8.44	11.3	14.8	18.5	21.0
13	3.57	4.11	5.01	5.89	7.04	9.30	12.3	16.0	19.8	22.4
14	4.07	4.66	5.63	6.57	7.79	10.2	13.3	17.1	21.1	23.7
15	4.60	5.23	6.26	7.26	8.55	11.0	14.3	18.2	22.3	25.0
16	5.14	5.81	6.91	7.96	9.31	11.9	15.3	19.4	23.5	26.3
17	5.70	6.41	7.56	8.67	10.1	12.8	16.3	20.5	24.8	27.6
18	6.26	7.01	8.23	9.39	10.9	13.7	17.3	21.6	26.0	28.9
19	6.84	7.63	8.91	10.1	11.7	14.6	18.3	22.7	27.2	30.1
20	7.43	8.26	9.59	10.9	12.4	15.5	19.3	23.8	28.4	31.4
30	13.8	15.0	16.8	18.5	20.6	24.5	29.3	34.8	40.3	43.8
40	20.7	22.2	24.4	26.5	29.1	33.7	39.3	45.6	51.8	55.8

[−], the first row provides confidence levels P [−], and the other rows and columns are values of χ^2 [−].

The number of bins n [−] must be selected subject to the constraint that $E_k > 5$ [−], i.e. in each bin at least 5 expected values are considered, corresponding to the 5 observed values. For $N \geq 25$ [−], and for equal-width bins, the numbers of bins may be estimated using the Scott's formula such that

$$n = 1.15 N^{1/3}, \tag{10.27}$$

where N [−] is the number of observed values.

10.10   Two Sample Statistical Estimation

The two sample statistical estimation test is used to compare two measured values, for each of which an average and standard deviation is available. This inferential test quantifies the range for the difference between the two averages. If $\bar{x}_1$ [−] and σ_1 [−], and $\bar{x}_2$ [−] and σ_2 [−] are the average and standard deviation of independent samples of size n_1 [−] and n_2 [−], respectively, an approximate $P = 1 - \alpha$ [−] confidence interval for $\mu_1 - \mu_2$ [−], i.e. the difference between the true means, is

$$(\bar{x}_1 - \bar{x}_2) - t_{\alpha/2}\sqrt{\frac{\sigma_1^2}{n_1} + \frac{\sigma_2^2}{n_2}} < \mu_1 - \mu_2 < (\bar{x}_1 - \bar{x}_2) + t_{\alpha/2}\sqrt{\frac{\sigma_1^2}{n_1} + \frac{\sigma_2^2}{n_2}}, \tag{10.28}$$

where $t_{\alpha/2}$ [−] is the t-value with degrees of freedom

$$\nu = \frac{\left(\frac{\sigma_1^2}{n_1} + \frac{\sigma_2^2}{n_2}\right)^2}{\frac{(\sigma_1^2/n_1)^2}{n_1-1} + \frac{(\sigma_2^2/n_2)^2}{n_2-1}}, \tag{10.29}$$

leaving an area $\alpha/2$ [−] to the right of a t probability distribution function [2].

10.11   Reporting Uncertainties

In general, the result of any measurement of a variable is reported as $\bar{x} \pm u_x$ [−], where $\bar{x}$ [−] is the average of a few measurements and u_x [−] is a measure of uncertainty. It is important to understand what $\pm u_x$ [−] means. In fact $\pm u_x$ [−] must be accompanied with a confidence level P [−] for the possible range of values for the measurement. So a complete way to report the measurement is

$$\bar{x} \pm u_x \ (P). \tag{10.30}$$

The value of $\pm u_x$ [–] is in fact equal to $u_x = z\sigma_x$ [–] or $u_x = t\sigma_x$ [–] if the normal or Student's t distributions are used, respectively, for a corresponding value of P [–]. Sometimes the fractional uncertainty is used instead of the uncertainty itself

$$\text{Fractional uncertainty} = \frac{u_x}{\bar{x}}. \tag{10.31}$$

Alternatively, percent uncertainty may be used, which is fractional uncertainty multiplied by 100.

10.12   Propagation of Uncertainties

An important topic is propagation of uncertainties. A real experiment often involves many different uncertainties that somehow must be combined to provide a total uncertainty for the outcome of the experiment.

Figure 10.4 shows the relationship between uncertainties u_x [–] and u_y [–], which belong to the independent variable x [–] and dependent variable $y = f(x)$ [–],

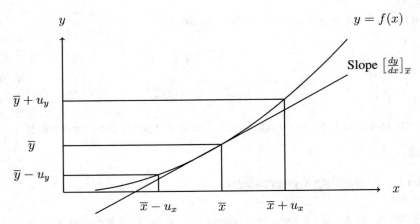

Fig. 10.4 The uncertainty in the dependent variable $\bar{y}$ [–], i.e. u_y [–], which is a function of the independent variable $\bar{x}$ [–] with uncertainty u_x [–]

which is a function of only a single independent variable x [–]. Using the Taylor series expansion, y [–] in the vicinity of $\overline{x}$ [–] can be written as

$$y = \overline{y} \pm u_y = f(\overline{x}) \pm \left(\left[\frac{dy}{dx} \right]_{\overline{x}} u_x + \frac{1}{2!} \left[\frac{d^2 y}{dx^2} \right]_{\overline{x}} u_x^2 + \dots \right), \tag{10.32}$$

where the higher order terms can be ignored if u_x [–] is small. This allows for estimating the uncertainty in y [–] such that

$$u_y = \left[\frac{dy}{dx} \right]_{\overline{x}} u_x. \tag{10.33}$$

Now consider a more general case where function $y = f(x_1, \dots, x_n)$ [–] is a function of many independent variables $(x_1, \dots, x_n)$ [(-,…,-)] with uncertainties $(u_{x_1}, \dots, u_{x_n})$ [(-,…,-)]. It can be shown that the uncertainty in y [–] at $\overline{y} = f(\overline{x}_1, \dots, \overline{x}_n)$ [–], i.e. u_y [–], near $(\overline{x}_1, \dots, \overline{x}_n)$ [(-,…,-)] can be given by [4]

$$u_y = \left[\sum_{i=1}^{n} \left(\left[\frac{\partial y}{\partial x_i} \right]_{\overline{x}_i} u_{x_i} \right)^2 \right]^{1/2}. \tag{10.34}$$

It must be noted that this formula only applies to the case where each of the two independent variables have zero correlation coefficient. Otherwise, a more complex formula can be used. For instance, consider the case where $y = f(x_1, x_2) = x_1 + x_2$ [–]. The uncertainty in y [–] at any $(\overline{x}_1, \overline{x}_2)$ [(-,-)] can be given as

$$u_y = \left[u_{x_1}^2 + u_{x_2}^2 \right]^{1/2}. \tag{10.35}$$

Problems

10.1 Four readings of a true voltage of 100 V yield 104, 103, 105, and 104 V. Find the average precision and accuracy of the measurement.

10.2 The fluctuating temperature of a turbulent gas stream has been measured using a number of $N = 36$ [–] samples, and the temperature has been found to be on average 20 °C with a standard deviation of 0.5 °C. Assuming that the temperature measurement is normally distributed, determine the confidence interval for the range of temperatures associated with $P = 0.9$ [–].

10.3 In an experiment, the following data are obtained by measuring a variable in a sample of $N = 10$ [–]. The data are 0.98, 1.07, 0.86, 1.16, 0.96, 0.68, 1.34, 1.04, 1.24, 0.86 [–]. Compute the average and standard deviation of the measurement. Assuming that the measured variable is distributed according to the Student's t distribution, determine the confidence interval for the range of data associated with $P = 0.9$ [–].

Table 10.4 Observations in an experiment with a sample size of $N = 40$ [–]

i	x_i [–]	i	x_i [–]	i	x_i [–]	i	x_i [–]
1	731	11	689	21	688	31	735
2	739	12	754	22	748	32	638
3	678	13	681	23	778	33	733
4	698	14	676	24	710	34	766
5	772	15	810	25	653	35	709
6	780	16	830	26	672	36	787
7	748	17	722	27	674	37	742
8	770	18	760	28	738	38	645
9	771	19	805	29	757	39	675
10	709	20	725	30	687	40	712

10.4 In an experiment, 12 measurements are taken: 12, 34, 22, 14, 22, 17, 24, 22, 18, 14, 18, and 12 [–]. Find the mean and standard deviation of this sample. According to the Chauvenet's criterion, would it be justified to reject 34 [–] from the data sample? Find the new mean and standard deviation if you reject 34 [–].

10.5 In an experiment, the following data for y [–] are obtained as independent variable x [–]. Values of x [–] are 1, 2, 3, 4, and 5 [–]. Values of y [–] are 1.2, 1.9, 3.2, 4.1, and 5.3 [–]. Using the method of least-squares fitting, find coefficients a_0 and a_1 [–] for a line fit $y_c = a_0 + a_1 x$ [–]. Calculate the standard error of the fit. Then using the Student's t distribution, estimate the random uncertainty associated with $P = 0.95$ [–] confidence level. Finally, calculate the correlation coefficient for this fit.

10.6 In an experiment, we make $N = 40$ [–] measurements $(x_1, \ldots, x_{40})$ [(-, \ldots, -)], and we obtain observed values according to Table 10.4. Use the Scott's formula to show that the number of bins for a Chi-squared test should be $n = 4$ [–]. Then find the average $\bar{x}$ [–] and standard deviation σ_x [–] of the measurements. Subsequently, define $n = 4$ [–] bins and group the data into the following ranges: $x \leq \bar{x} - \sigma_x, \bar{x} - \sigma_x < x \leq \bar{x}, \bar{x} < x \leq \bar{x} + \sigma_x$, and $\bar{x} + \sigma_x < x$ [–]. Count the number of observations in each bin. These are values of O_k [–]. Compute the bin width using $\frac{x_{max} - x_{min}}{n}$ [–], and verify that the bin width is in the order of the sample standard deviation σ_x [–]. We expect that the observed values follow a normal distribution. Follow the table for the normal distribution to calculate the probability of the expected value to fall in the specified ranges of each bin, i.e. $\int_k f(x)dx$ [–] with x [–] varying according to bin k [–] lower and upper limits. These probabilities provide values of $E_k = N \int_k f(x)dx$ [–]. This allows you to calculate the Chi-squared as

$$\chi^2 = \sum_{k=1}^{n} \frac{(O_k - E_k)^2}{E_k}, \qquad (10.36)$$

with degrees of freedom $\nu = n - 3$ [–]. Use this value of χ^2 [–] to provide an estimate for the probability or confidence level P [–], for which the observed values follow a normal distribution. Is this probability high enough to indicate that our assumption is probably correct?

10.7 Compute the percentage uncertainty for the following measurement: $x = 3.3 \pm 1.4$ mm.

10.8 Use the theory of propagation of uncertainty to compute u_y [–] near $\overline{x} = 2$ [–] if $y = x^2$ [–] and $u_x = 0.1$ [–].

10.9 The voltage and current for an electrical heater are $E = 110 \pm 3$ V and $I = 20.0 \pm 0.7$ A. Use theory of propagation of uncertainty to compute the value and the uncertainty for the power consumption of the heater $P = EI$ [W].

References

1. Barlow R J (1993) Statistics: a guide to the use of statistical methods in the physical sciences, 1st edn. Wiley, New York
2. Walpole R E, Myers, R H, Myers S L et al (2002) Probability & statistics for engineers & scientists, 7th edn. Prentice Hall, Upper Saddle River
3. Nambiar M K, Robe F, Seguin A M et al (2020) Diurnal and seasonal variation of area-fugitive methane advective flux from an open-pit mining facility in northern Canada using WRF. Atmosphere-Basel 11:1227
4. Ku H H (1966) Notes on the use of propagation of error formulas. J Res Nat Bur Stand Sec C: Eng Inst 70C:263–273

Chapter 11
In Situ Techniques

Abstract This chapter introduces the in situ measurement techniques, as approaches that rely on the physical presence of a sensor at the location of interest in the fluid, at which properties are to be measured. A selection of sensors are discussed, involving U-tube manometers, strain gauge pressure transducers, electrical resistance thermometry, thermoelectric temperature measurement, hot wire anemometry, pitot tubes, rotameters, and balloons. As part of this chapter, the equations describing the bridge circuits and the Cartesian coordinate system rotation transformation are provided.

11.1   Overview

In situ techniques for measuring turbulence are intrusive, meaning that the measuring probe will be placed at the location where the flow is to be measured. This may likely influence, or disturb, the flow. However, since the probe is placed at the very location of the flow to be measured, the technique may be more accurate than non-intrusive techniques. In this chapter measurements of flow pressure, temperature, and velocity will be discussed. The particular focus will be on methods suitable for the measurement of the atmospheric boundary layer [1].

11.2   U-Tube Manometer

U-tube manometers are the simplest devices to measure the pressure of a fluid line. As shown in Fig. 11.1, the manometer is filled with manometer liquid fluid. One side is connected to a pressure line to be measured (P) [Pa], while the other side is open to the atmosphere with atmospheric pressure P_a [Pa]. The manometer rule states that pressures P_1 and P_2 [Pa] must be equal, i.e. $P_1 = P_2$ [Pa]. This provides

Fig. 11.1 Schematic of a
U-tube manometer

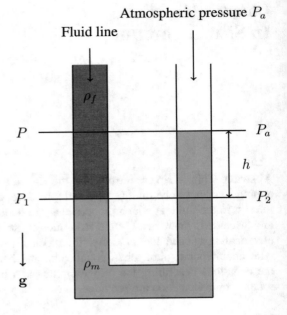

an estimate for the unknown pressure of the fluid P [Pa], such that

$$P_1 = P + \rho_f gh, \tag{11.1}$$

$$P_2 = P_a + \rho_m gh, \tag{11.2}$$

$$P = P_a + (\rho_m - \rho_f)gh. \tag{11.3}$$

Note that here ρ_f [kg m^{-3}] is the density of the fluid, whose pressure is to be measured, and ρ_m [kg m^{-3}] is the density of the manometer fluid. In this derivation it is assumed that the pressure change due to the column of the atmosphere is negligible with respect to the other pressure changes.

11.3   Strain Gauge Pressure Transducers

Strain gauge pressure transducers are widely used for narrow-span pressure and differential pressure measurements. As shown in Fig. 11.2, the strain gauge is used to measure the displacement of an elastic diaphragm due to a differential pressure. Strain gauges are transducers that experience a change of electric resistance when they are strained. A calibration experiment can help relate the change in resistance to the pressure differential. The change in resistance may be measured using a bridge circuit, such as the Wheatstone bridge.

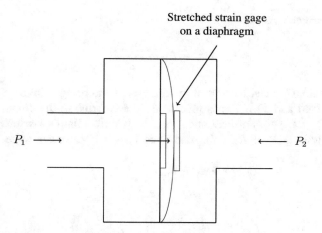

Fig. 11.2 Strain gauge pressure transducer to measure a pressure differential

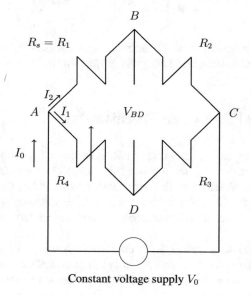

Constant voltage supply V_0

Fig. 11.3 Schematic of a constant voltage Wheatstone bridge circuit for a strain gauge pressure transducer

Figure 11.3 shows the schematic of a constant voltage Wheatstone bridge circuit for a strain gauge pressure transducer. The bridge is designed to measure resistance $R_s = R_1$ [Ω], having known resistances R_2 and R_3 [Ω], by adjusting the variable resistance R_4 [Ω] until the voltage measured between points B and D, i.e. V_{BD} [V], becomes zero. The bridge is known to be balanced when $V_{BD} = 0$ V, where R_s [Ω]

can be determined via [2]

$$R_s = R_1 = \frac{R_2 R_4}{R_3}.$$ (11.4)

This relationship can be proven by analysis of the bridge circuit. Suppose that V_0 [V] is fixed and V_{BD} [V] is measured. The currents in the circuit are related by $I_0 = I_1 + I_2$ [A]. Further, since $V_{BD} = 0$ V, the Ohm's law provides $V_0 = I_1(R_4 + R_3) = I_2(R_1 + R_2)$ [V]. Also since $V_{BD} = 0$ V, we can write

$$V_{BD} = V_{BA} + V_{AD},$$
$$0 = -I_2 R_1 + I_1 R_4,$$
$$0 = -\frac{V_0}{R_1 + R_2} R_1 + \frac{V_0}{R_4 + R_3} R_4,$$
$$0 = V_0 \left[-\frac{R_1}{R_1 + R_2} + \frac{R_4}{R_4 + R_3} \right].$$ (11.5)

This directly results in the relationship for resistances noted above. The derivation for the constant current bridge circuits is very similar.

11.4   Electrical Resistance Thermometry

Thermometry refers to the techniques that measure temperature. The principle behind electrical resistance thermometry is that the electrical resistance of a conductor or semiconductor varies with temperature, which can be exploited to measure temperature. Common techniques include Resistance Temperature Detectors (RTDs) that use conductor materials, while thermistors use semiconductor materials.

Shown in Fig. 11.4, a typical RTD consists of a wire coil sensor, sheath for protection, Wheatstone bridge, and a voltage display instrument. The sensor part of an RTD is a conductor element that exhibits a resistance temperature relationship given by

$$R_{RTD} = R_0 \left(1 + \alpha T + \beta T^2 + \dots \right),$$ (11.6)

Fig. 11.4 Schematic of a Resistance Temperature Detector (RTD)

Sheath for protection

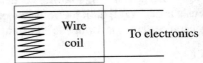

Wire coil

To electronics

where R_0 [Ω] is the electrical resistance at some reference temperature $T_0 = 0\,°C$, and α [$°C^{-1}$], β [$°C^{-2}$], ... are the empirical constants. Using Taylor expansion, for a limited range of temperatures, we can write the following approximation to relate the change in resistance to the change in temperature

$$\frac{\Delta R}{R_0} = \frac{R_{RTD} - R_0}{R_0} = \alpha(T - T_0). \tag{11.7}$$

Therefore, measuring the change in temperature would be possible by measuring the change in resistance. The resistance can be measured using a bridge circuit, such as the constant voltage bridge circuit. Figure 11.5 shows the schematic of a constant voltage Wheatstone bridge circuit for an RTD. The bridge is designed to measure resistance $R_{RTD} = R_1$ [Ω], having known resistances R_2 and R_3 [Ω], by adjusting the variable resistance R_4 [Ω] until the voltage measured between points B and D, i.e. V_{BD} [V], becomes zero. The bridge is known to be balanced when $V_{BD} = 0\,V$, where R_{RTD} [Ω] can be determined via [2]

$$R_{RTD} = R_1 = \frac{R_2 R_4}{R_3}. \tag{11.8}$$

The most common material used for RTDs is platinum ($\alpha = 0.00392\,°C^{-1}$). RTDs may be used for the measurement of temperatures ranging from cryogenic temperatures to approximately $650\,°C$. Using RTDs, an uncertainty in temperature measurement as low as $\pm 0.005\,°C$ is possible. Although accurate, RTDs can be expensive.

Fig. 11.5 Schematic of a constant voltage Wheatstone bridge circuit for an RTD

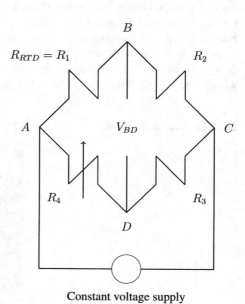

Constant voltage supply

The thermistor is fabricated using semiconductor materials (e.g. oxides of nickel, cobalt, sulphides of iron, aluminium, or copper). Unlike conductors, semiconductors exhibit a decrease in resistance with an increase in temperature via

$$R_T = R_0 \exp\left[\beta\left(\frac{1}{T} - \frac{1}{T_0}\right)\right], \tag{11.9}$$

where R_0 [Ω] is the resistance for a reference temperature T_0 [K], and β [K] is an experimental constant ranging between 3500 and 4600 K. Thermistors are generally used when high sensitivity or fast response times are required. Semiconductors are subject to deterioration at high temperatures, so their use is limited to temperatures only as high as 300 °C. Using thermistors, an uncertainty in temperature measurement as low as ±0.01 °C is possible. Again the resistance of a thermistor can be measured using a bridge circuit.

11.5   Thermoelectric Temperature Measurement

The thermoelectric phenomena include Seebeck, Peltier, and Thompson effects. The Seebeck effect occurs when two dissimilar metals are joined together as shown in Fig. 11.6. When the junctions A and B are maintained at two different temperatures, say hot T_A [K] and cold T_B [K] temperatures, then a voltage V [V] appears between the metals.

The voltage is a function of the difference in junction temperatures and the metal materials used. The Seebeck effect refers to the case when there is no current in the circuit, so that the circuit is essentially open. Thermocouples exploit the Seebeck effect such that if the temperature of junction B is known, then the temperature of junction A can be known by measuring V [V]. For instance the cold junction B may be in thermal equilibrium with an ice-water bath at a known pressure, in which case T_B [K] would be known. At least two metals must be used to construct

Fig. 11.6 The Seebeck effect   Hot junction A Cold junction B

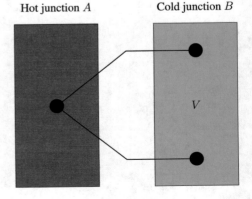

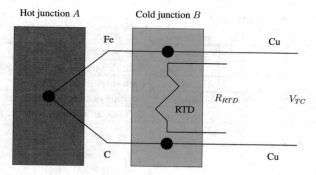

Fig. 11.7 A thermocouple circuit with electronically compensated reference temperature using an RTD

a thermocouple circuit. The two metals are typically welded for ideal electrical contact. The law of intermediate metals states that the net voltage in a thermocouple circuit, which is composed of any number of dissimilar materials, is not affected if the dissimilar metals are connected at the same temperature. In other words, the voltage will only be affected if the dissimilar metals are connected at different temperatures.

Most modern thermocouple systems do not require a fixed reference temperature, such as the temperature of an ice-water bath, at junction B. Alternatively, the temperature of junction B can be measured by other means, such as the use of an RTD sensor. In such cases, it is stated that the reference temperature is electronically compensated or measured. Figure 11.7 shows such a circuit where an RTD is used for electronic compensation. Here the thermocouple is made of Iron (Fe) and Constantan (C), while the rest of the wiring in the circuit is made of Copper (Cu). In this circuit the RTD is positioned inside the cold junction B, so it is at the same temperature. The temperature of junction A can be determined by measuring the resistance R_{RTD} [Ω] (using a bridge circuit) and voltage V_{TC} [V].

Table 11.1 shows the commonly used thermocouples. For each thermocouple type, standard calibration curves are available that provide a polynomial relationship between the temperature and the voltage such that

$$T = a_0 + a_1 V + a_2 V^2 + \cdots + a_9 V^9, \tag{11.10}$$

where T [K] is the temperature, V [V] is the voltage assuming that reference junction is maintained at some temperature, and a_0 [K], a_1 [K V^{-1}], a_2 [K V^{-2}], $\ldots$ are the polynomial coefficients. The polynomial curves exhibit close-to-linear behaviour over finite ranges of temperature. Unlike RTDs and thermistors, thermocouples exhibit higher uncertainty in temperature measurement as high as $\pm 2\,°C$, so it is important to calibrate them properly.

Table 11.1 Thermocouple types commonly used

Type	Positive (+)	Negative (−)	Application
E	Chromel	Constantan	Highest sensitivity (<1000 °C)
J	Iron	Constantan	Nonoxidizing environment (<760 °C)
K	Chromel	Alumel	High temperature (<1372 °C)
S	Platinum–10% Rhodium	Platinum	Durability and high temperature (<1768 °C)
T	Copper	Constantan	Vacuum environment (<400 °C)

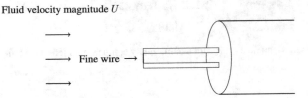

Fig. 11.8 Schematic of a hot wire anemometer

11.6   Hot Wire Anemometry (HWA)

Hot wire anemometers use a fine wire (on the order of several micrometres) electrically heated to some temperature above the ambient. Air flowing past the wire removes heat from the wire. As the electrical resistance of most metals is dependent upon the temperature of the metal (tungsten is a popular choice for hot wires, but platinum or iridium is also used), a relationship can be obtained between the resistance, heat removed from the wire, and ultimately the flow velocity.

Several ways of implementing this exist, and hot wire devices can be further classified as Constant Current Anemometer (CCA), Constant Voltage Anemometer (CVA), and Constant Temperature Anemometer (CTA). The voltage output from these anemometers is thus the result of some sort of circuit within the device trying to maintain a specific variable (current, voltage, or temperature) constant, following Ohm's law.

Additionally, Pulse-Width Modulation (PWM) anemometers are used, wherein the velocity is inferred by the time length of a repeating pulse of current that brings the wire up to a specified resistance and then stops until a threshold *floor* is reached, at which time the pulse is sent again.

Hot wire anemometers, while extremely delicate, have extremely high frequency response and fine spatial resolution compared to other measurement methods, and as such are almost universally employed for the detailed study of turbulent flows, or any flow in which rapid velocity fluctuations are of interest. For special applications, frequencies as high as 1 MHz can be achieved [2], while typical frequencies are around 1 kHz [3].

Figure 11.8 shows the schematic of a hot wire anemometer. The rate of heat transferred between the hot wire maintained at temperature T_s [K] and the cooler fluid maintained at temperature T_f [K] is proportional to the local convective heat transfer coefficient h_w [W m^{-2} K^{-1}] and the heat transfer area A_w [m^2] of the wire. According to King's relation for hot wire anemometer, the heat transfer rate can be written as

$$q_w = h_w A_w (T_s - T_f) = \left(a + bU^{0.5}\right) A_w (T_s - T_f),\qquad (11.11)$$

where U [m s^{-1}] is the magnitude of the fluid velocity vector normal to the wire axis, and a and b are the constants obtained by calibration. At the same time, the heat transfer rate can be computed using the resistance of the fine wire R_w [Ω] and the current through it I_w [A] via

$$q_w = R_w I_w^2 = R_0 \left[1 + \alpha(T_s - T_0)\right] I_w^2,\qquad (11.12)$$

where R_0 [Ω] is the resistance of the wire at a reference temperature T_0 [K], and α [K^{-1}] is a constant. So quantification of the flow velocity magnitude U [m s^{-1}] eventually falls upon measurement of the wire resistance and the current. The resistance, and subsequently the current, can be measured using a bridge circuit, such as the constant current bridge circuit.

Figure 11.9 shows the schematic of a constant current Wheatstone bridge circuit for a hot wire anemometer. The bridge is designed to measure resistance $R_w = R_1$ [Ω], having known resistances R_2 and R_3 [Ω], by adjusting the variable resistance R_4 [Ω] until the voltage measured between points B and D, i.e. V_{BD} [V], becomes zero. The bridge is known to be balanced when $V_{BD} = 0$ V, where R_w [Ω] can be

Fig. 11.9 Schematic of a constant current Wheatstone bridge circuit for a hot wire anemometer

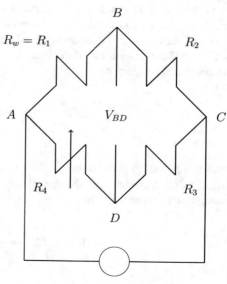

Constant current supply

determined via [2]

$$R_w = R_1 = \frac{R_2 R_4}{R_3}.$$ (11.13)

In practice, multiple hot wires may be placed in different orientations, so the desired components of the flow velocity vector can be measured. Alternative to hot wire anemometers, hot film anemometers may be used in environments where the hot wire may be damaged, such as where measuring water flow velocities. The principle of operation for hot film anemometers is similar to that of hot wire anemometers.

11.7   Pitot Tube

A pitot–static tube, which is a pitot tube with two ports, pitot and static, is normally used in measuring the airspeed of aircraft and atmospheric turbulence [4]. The pitot port measures the dynamic pressure of the open mouth of a tube with the pointed head facing wind, and the static port measures the static pressure from small holes along the side on that tube. The pitot tube is connected to a tail so that it always makes the tube's head to face the wind. Additionally, the tube is heated to prevent rime ice formation on the tube. There are two lines from the tube down to the devices to measure the difference in pressure of the two lines. The measurement devices can be manometers, pressure transducers, or analog chart recorders.

Figure 11.10 shows the schematic of a pitot–static tube, consisting of a channel to which the piezometer and pitot tubes are connected. Note that in practice the channel is very narrow, so it may be assumed that the pressure across the channel does not vary (except for a thin layer near the channel wall where the flow becomes stagnant). For a streamline from point D (dynamic) to point S (static), the Bernoulli equation can be written as

$$P_D + \frac{\rho V_D^2}{2} = P_S + \underbrace{\frac{\rho V_S^2}{2}}_{=0},$$ (11.14)

where ρ [kg m^{-3}] is the density of the fluid in the tubes. The static velocity $V_S = 0$ m s^{-1} due to the fact that the flow facing the tip of the pitot tube is stagnant. This provides an estimate for the dynamic velocity V_D [m s^{-1}] given by

$$V_D = \sqrt{2 \frac{P_S - P_D}{\rho}} = \sqrt{2gh},$$ (11.15)

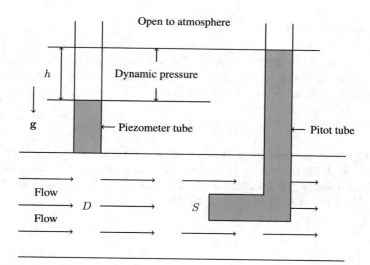

Fig. 11.10 Schematic of a pitot–static tube

where g [m s^{-2}] is the gravitational acceleration. Small geometric errors in the static pressure tap (e.g. machining details) may result in significant static pressure measurement errors. The static pressure recorded by a pitot–static tube is usually less than the true static pressure. This is due to the increase of the fluid velocity near the tube and channel.

Pitot tubes can also be used to measure velocity in compressible flows [5]. For the ideal compressible gas flows (i.e. non-viscous flows) with no heat transfer, the Euler Eq. 2.10 can be used to show that

$$\frac{P_S}{P_D} = \left(1 + \frac{k-1}{2}M_D^2\right)^{\frac{k}{k-1}}, \qquad (11.16)$$

where $k = c_p/c_v$ [–] is the ratio of specific heats in the gas and M_D is the Mach number given by

$$M_D = \frac{V_D}{\sqrt{kRT/M}}, \qquad (11.17)$$

in which R [JK^{-1} mol^{-1}] is the gas constant, M [kgmol^{-1}] is the gas molar mass, and T [K] is absolute temperature. This equation can be expressed in terms of the velocity using the stagnation temperature T_S [K] measured by a thermal sensor near the stagnation point. Therefore, the dynamic velocity can be given satisfying

$$\frac{V_D^2}{2} = c_p T_S \left[1 - \left(\frac{P_D}{P_S}\right)^{\frac{k-1}{k}}\right]. \qquad (11.18)$$

11.8   Rotameters

Rotameters are widely used for liquid and gas flow rate measurements. Schematic of a rotameter is shown in Fig. 11.11. In this technique a float is suspended by a balance of three forces in the vertical direction

$$F_D + F_B = W, \tag{11.19}$$

where F_D [N] is the drag force on the float, F_B [N] is the buoyancy force on the float, and W [N] is the weight of the float. The drag force can be given using the coefficient of drag C_D [–], frontal area of the float A_f [m^2], density of the fluid ρ_f [kg m^{-3}], and the velocity of the fluid V_f [m s^{-1}]

$$F_D = C_D A_f \frac{\rho_f V_f^2}{2}. \tag{11.20}$$

Knowing F_D [N] and subsequently V_f [m s^{-1}], it can be shown that the volumetric flow rate can be obtained as $Q = V_f(A_z - A_f)$ [m^3 s^{-1}], where A_z [m^2] is the annulus area (a function of height z [m]). For a given fluid velocity, V_f [m s^{-1}] will remain constant. Thus higher flow rates imply a rise of the float to height z [m], so that the flow rate can be determined by the position of the float z. Rotameters are calibrated for flow rate given a specific fluid type (e.g. air, oxygen, water).

Fig. 11.11 Schematic of a rotameter

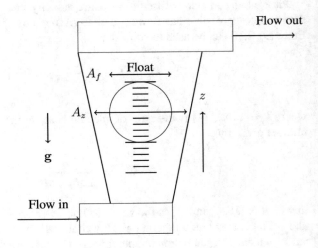

11.9 Balloons

Balloons are among the earliest in situ techniques that have been used to measure atmospheric turbulence. Balloons use either hot air or a light gas, employing helium or other light gases, to levitate a suite of sensors in the atmosphere by the buoyancy force. Balloons can either be launched once without returning, the so-called radio sondes, or they can be tethered for multiple use [2, 6]. Figure 11.12 shows a tethered-balloon system controlled using two ropes [7].

Airborne systems are increasingly being used for atmospheric measurements [8, 9] although recently their use is being regulated more restrictively. For instance, rotary or fixed-wing drones are not permitted to fly in complex environments such as busy urban areas and airports. On the other hand, tethered-balloon-based atmospheric measurement techniques have been used widely for obtaining the turbulence structure as well as the mean vertical profiles of the atmospheric boundary layer meteorological variables in complex environments [10]. One of the main advantages of a tethered-balloon system is its ability to profile a significant portion of the planetary boundary layer starting from the surface, which is not possible or economical by ground-based or aircraft-based atmospheric measurement techniques [11]. The use of ultrasonic anemometers in tethered balloons has been

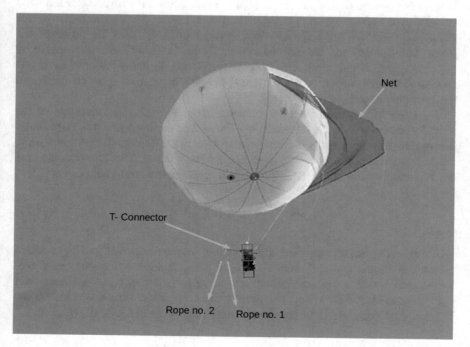

Fig. 11.12 A tethered-balloon system for atmospheric measurements controlled using two ropes [7]

reported in many studies [12]. In comparison, one of the disadvantages of pitot tubes is their inability to measure the low wind speeds. So they require a fast flying probe that cannot fly in a complex environment for safety and logistic reasons. Ultrasonic anemometers, on the other hand, are popular because of their continuous measurement characteristics, high accuracy, and their ability to be levitated to measure low velocities.

Tethered-balloon-borne atmospheric turbulence measurements have a long history of observations over the land [13] and sea [14] to measure fluxes of heat and moisture at heights up to a few hundred metres. The most notable tethered-balloon systems deployed collected data in campaigns in the late 1960s and 1970s including the Barbados Oceanographic and Meteorological Experiment (BOMEX) [15–17], the Joint Air–Sea Interaction (JASIN) experiment [18], and the Global Atmospheric Research Programme (GARP) Atlantic Tropical Experiment (GATE) [19]. In BOMEX a tethered-balloon system was operated from the deck, which measured temperature, wind, and humidity continuously, at different levels in the range of 0–600 m in the ocean area north and east of the Island of Barbados. In JASIN tethered balloons were used to measure the structure of Atmospheric Boundary Layer (ABL) to understand the air–sea interaction in the North Atlantic. In the recent past, tethered-balloon systems have been used in Boundary Layer Late Afternoon and Sunset Turbulence (BLLAST) field campaign that was conducted in southern France [20]. Canut et al. [12] used an ultrasonic anemometer mounted on a tethered-balloon system for turbulent flux and variance measurements. Egerer et al. [11] used the BELUGA (Balloon-bornE moduLar Utility for profilinG the lower Atmosphere) tethered-balloon system for profiling the lower atmosphere by turbulence and radiation measurements in the Arctic. Tethered balloons have also been used to perform earth surface thermal imaging in complex open-pit mines and the surrounding complex terrain [21].

An important measurement for tethered balloons is the measurement of the altitude. As reported in the literature, the Global Positioning System (GPS)-sensor-derived altitude can result in uncertainties up to 50 m as stated by [22]. Alternatively, the hypsometric equation can be used to calculate altitude. This equation uses the atmospheric pressure and accounts for atmospheric temperature changes via [23]

$$z_2 - z_1 \approx a\overline{T_v} \ln\left(\frac{P_1}{P_2}\right), \tag{11.21}$$

where z_1 and z_2 [m] represent the altitudes (in metres) corresponding to the recorded pressure measurements, P_1 and P_2 [Pa], $\overline{T_v}$ [K] represents the average virtual temperature between the two altitudes (z_1 and z_2 [m]), and a [m K^{-1}] is a constant equivalent to 29.3 m K^{-1} [23]. The uncertainty for altitude using this equation can be quantified using theory of error propagation, Eq. 10.34 in Chap. 10, via [24].

$$\Delta z_2 = \sqrt{\left(\frac{\partial z_2}{\partial \overline{T_v}}\right)^2 \Delta \overline{T_v}^2 + \left(\frac{\partial z_2}{\partial P_2}\right)^2 \Delta P_2^2}, \tag{11.22}$$

$$\frac{\partial z_2}{\partial \overline{T_v}} = a \ln\left(\frac{P_1}{P_2}\right),$$ (11.23)

$$\frac{\partial z_2}{\partial P_2} = a\overline{T_v}\left(\frac{-1}{P_2}\right),$$ (11.24)

where $\Delta \overline{T_v}$ [K] and ΔP_2 [Pa] are the uncertainties for virtual temperature and pressure measurements using the balloon weather instruments. Byerlay et al. [21] quantified the uncertainty for a tethered balloon to be approximately equal to 1.2 m, which is much lower than typical uncertainty for GPS-derived altitudes.

Since balloons typically rotate about three axes once levitated in air, their measurements of wind velocity vector components should be referenced to a frame with fixed coordinate axes directions along east, north, and normal to the earth surface using rotation matrices. This is made possible by use of the heading, roll, and pitch angles measured by sensors on board. Suppose that the sensor has a coordinate system in which the $+x$ axis is from local north N_T of the sensor to local south S_T, the $+y$ axis is from local east E_T to local west W_T, and the $+z$ axis is downward, i.e. towards the earth surface. This local coordinate system is *right handed*. Likewise, the velocities measured by the sensor are positive along these axes. Let the sensor's coordinate system be shown by x_T, y_T, and z_T and the corresponding velocities be shown by U_T, V_T, and W_T [m s^{-1}]. The sensor measures heading h [°], positive clockwise with respect to magnetic north N_M. It measures the pitch angle p [°], which is a positive downward rotation of x_T about y_T, and the roll angle r [°], which is a positive downward rotation of y_T about x_T. The goal is to transform this coordinate system, by means of rotation matrices, to align with a reference frame of the earth with x_F pointing from west to east, y_F pointing from south to north, and z_F pointing away upward from the surface of the earth. This coordinate system is also *right handed*. The resulting velocity transformation will provide U_F, V_F, and W_F [m s^{-1}] in the final coordinate system.

As depicted in Fig. 11.13, the first rotation should be about the local y_T axis by an angle $\gamma = p$ [°]. This transformation results in an intermediate coordinate system x_1, y_1, and z_1 so that the x_1 axis will be aligned with the horizon, i.e. parallel to the earth surface. Note that the figure is viewed normal to $y_T = y_1$ and that in this coordinate system z_1 is still not yet normal to the earth surface and that y_1 is still

Fig. 11.13 Rotation about y_T by $\gamma = p$ [°]; figure viewed normal to the $y_T = y_1$ axis

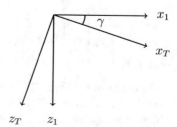

Fig. 11.14 Rotation about x_1 by $\eta = r$ [°]; figure viewed normal to the $x_1 = x_2$ axis

Fig. 11.15 Rotation about z_2 by $\alpha = (\delta + h + 90)\%360$ [°]; figure viewed normal to the $z_2 = z_3$ axis

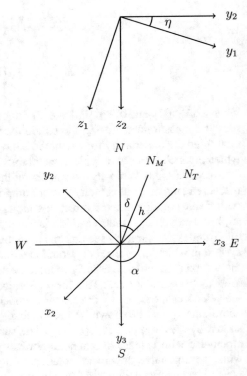

not yet aligned with the horizon. This transformation is given by

$$
\begin{bmatrix} x_1 \\ y_1 \\ z_1 \end{bmatrix} = \begin{bmatrix} \cos(\gamma) & 0 & \sin(\gamma) \\ 0 & 1 & 0 \\ -\sin(\gamma) & 0 & \cos(\gamma) \end{bmatrix} \begin{bmatrix} x_T \\ y_T \\ z_T \end{bmatrix} = R_{y,\gamma} \begin{bmatrix} x_T \\ y_T \\ z_T \end{bmatrix}.
\tag{11.25}
$$

As depicted in Fig. 11.14, the second rotation should be about x_1 axis by an angle $\eta = r$ [°]. This transformation results in another intermediate coordinate system x_2, y_2, and z_2 so that now the y_2 axis will be aligned with the horizon, i.e. parallel to the earth surface. Note that the figure is viewed normal to $x_1 = x_2$ and that in this coordinate system z_2 is normal to the earth surface. This transformation is given by

$$
\begin{bmatrix} x_2 \\ y_2 \\ z_2 \end{bmatrix} = \begin{bmatrix} 1 & 0 & 0 \\ 0 & \cos(\eta) & -\sin(\eta) \\ 0 & \sin(\eta) & \cos(\eta) \end{bmatrix} \begin{bmatrix} x_1 \\ y_1 \\ z_1 \end{bmatrix} = R_{x,\eta} \begin{bmatrix} x_1 \\ y_1 \\ z_1 \end{bmatrix}.
\tag{11.26}
$$

As depicted in Fig. 11.15, the third rotation should be about z_2 axis by an angle $\alpha = (\delta + h + 90)\%360$ [°], where δ [°] is the magnetic declination of the earth, which is dependent on a specific latitude and longitude. For example, in northern Alberta, Canada, $\delta = 13.5°$. Here the modulus with 360 is taken since the heading

Fig. 11.16 Rotation about x_3 by 180°; figure viewed normal to the z_3 axis

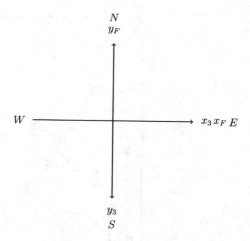

angle can vary from 0 to 360°. This transformation results in another intermediate coordinate system x_3, y_3, and z_3 so that now the y_3 axis will be aligned from north to south and the x_3 axis will be aligned from west to east. Note that the figure is viewed normal to $z_2 = z_3$. This transformation is given by

$$
\begin{bmatrix} x_3 \\ y_3 \\ z_3 \end{bmatrix} = \begin{bmatrix} \cos(\alpha) & -\sin(\alpha) & 0 \\ \sin(\alpha) & \cos(\alpha) & 0 \\ 0 & 0 & 1 \end{bmatrix} \begin{bmatrix} x_2 \\ y_2 \\ z_2 \end{bmatrix} = R_{z,\alpha} \begin{bmatrix} x_2 \\ y_2 \\ z_2 \end{bmatrix}.
\tag{11.27}
$$

As depicted in Fig. 11.16, the fourth and final rotation should be about x_3 axis by an angle 180°. This transformation results in the final coordinate system with x_F, y_F, and z_F axes, which point west to east, south to north, and normal upward from the earth surface, respectively. This transformation is given by

$$
\begin{bmatrix} x_F \\ y_F \\ z_F \end{bmatrix} = \begin{bmatrix} 1 & 0 & 0 \\ 0 & \cos(180°) & -\sin(180°) \\ 0 & \sin(180°) & \cos(180°) \end{bmatrix} \begin{bmatrix} x_3 \\ y_3 \\ z_3 \end{bmatrix} = R_{x,180°} \begin{bmatrix} x_3 \\ y_3 \\ z_3 \end{bmatrix}.
\tag{11.28}
$$

In compressed form, the entire coordinate transformation can be shown as follows. This immediately implies a similar transformation for the measured velocities by the sensor.

$$
\begin{bmatrix} x_F \\ y_F \\ z_F \end{bmatrix} = R_{x,180°} R_{z,\alpha} R_{x,\eta} R_{y,\gamma} \begin{bmatrix} x_T \\ y_T \\ z_T \end{bmatrix},
\tag{11.29}
$$

$$
\begin{bmatrix} U_F \\ V_F \\ W_F \end{bmatrix} = R_{x,180°} R_{z,\alpha} R_{x,\eta} R_{y,\gamma} \begin{bmatrix} U_T \\ V_T \\ W_T \end{bmatrix}.
\tag{11.30}
$$

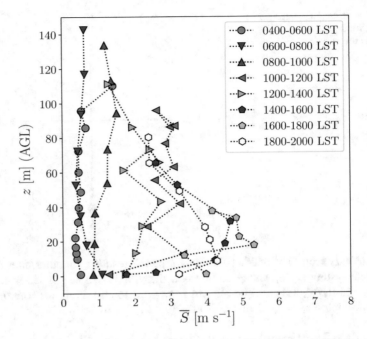

Fig. 11.17 Vertical profiles of wind speed measured in a mine field in northern Canada at different diurnal times in Local Standard Time (LST) using a tethered balloon [25]

Figures 11.17 and 11.18 show vertical profiles of wind speed and potential temperature measured using a tethered balloon in a mine field located in northern Canada at different diurnal times [7, 25]. Potential temperature is defined using Eq. 15.31 in Chap. 15. The wind speed profiles show calm atmospheric conditions in the early morning hours, while a low-level jet, to be introduced in Chap. 18, develops in the afternoon hours. The potential temperature profiles indicate thermally stable conditions in the early morning hours and thermally unstable conditions in the afternoon hours, with the concept of thermal stability to be defined in Chap. 18.

Problems

11.1 Consider a strain gauge pressure transducer used to measure the pressure differential in a system ranging from 1000 Pa to 10,000 Pa. A Wheatstone bridge with constant voltage $V_0 = 10$ V is used. The strain gauge is located at $R_s = R_1$ [Ω]. Initially, all arms of the bridge have resistances equal to $R_1 = R_2 = R_3 = R_4 = 100\,\Omega$. However, the pressure differential ΔP [Pa] results in a change in $R_s = R_1$ [Ω] according to $\Delta P = 1000 + 180(R_1 - 100)$ [Pa]. Note that in this bridge circuit resistance R_4 [Ω] is not adjusted to zero V_{BD} [V]. In an experiment with a non-zero ΔP [Pa], it is found that $V_{BD} = 0.65$ V. Find ΔP [Pa].

11.2 A thermistor with resistance R_T [Ω] is placed in series with another resistor $R_1 = 120$ kΩ and connected to a constant voltage supply of $E_i = 1.6$ V. For a

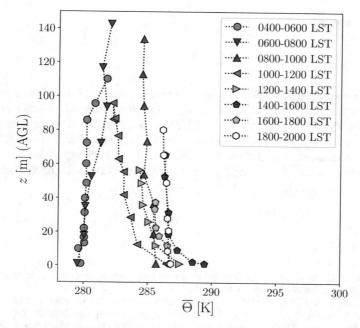

Fig. 11.18 Vertical profiles of potential temperature measured in a mine field in northern Canada at different diurnal times in Local Standard Time (LST) using a tethered balloon [25]

known temperature of $T = 130\,°C$, it is noted that the voltage across the known resistance is $E_1 = 1.54\,V$. Find the material constant β and its units for the thermistor using this circuit. Assume that the resistance at a reference temperature of $T_0 = 25\,°C$ is $R_0 = 50\,k\Omega$. Does this value of β fall within the typical range for thermistors?

11.3 An engineer has a thermocouple circuit, which consists of an RTD with a bridge circuit and a circuit to measure the thermocouple voltage, for measuring temperature using a single thermocouple. Instead of a single temperature, he wishes to measure the difference in temperature between two objects. Can he design a circuit in an ingenious way to make such a measurement with the existing circuits? Does he need the RTD sensor with the bridge circuit? Assume he has enough wiring of any kind and can weld, insulate, or glue wires together.

11.4 In hot wire anemometry a constant current Wheatstone bridge circuit may be used. As shown in Fig. 11.9, the bridge is designed to measure resistance $R_w = R_1$ [Ω] of the anemometer, having known resistances R_2 and R_3 [Ω], by adjusting the variable resistance R_4 [Ω] until the voltage measured between points B and D, i.e. V_{BD} [V], becomes zero. When the bridge is balanced ($V_{BD} = 0\,V$), show that R_w

[Ω] can be given by

$$R_w = R_1 = \frac{R_2 R_4}{R_3}. \tag{11.31}$$

11.5 A meteorologist wishes to quantify the uncertainty for the altitude measurement of a tethered balloon. Suppose that pressure measurements at altitudes z_2 and z_1 [m] are $P_2 = 100\,\text{kPa}$ and $P_1 = 101.3\,\text{kPa}$, respectively. Also suppose that the average virtual temperature measurement is $\overline{T_v} = 300\,\text{K}$. The uncertainty of the pressure measurement is $\Delta P_2 = 0.01\,\text{kPa}$, and the uncertainty of the temperature measurement is $\Delta \overline{T_v} = 2\,\text{K}$. Help the meteorologist calculate the uncertainty of the altitude measurement, i.e. Δz_2 [m].

References

1. Kaimal J C, Finnigan J J (1994) Atmospheric boundary layer flows: their structure and measurement. Oxford University Press, Oxford
2. Sumińska O A (2008) Application of constant temperature anemometer for balloon-borne stratospheric turbulence soundings. Leibniz Institute for Atmospheric Physics, Kühlungsborn
3. Hussein H J, Capp S P, George W K (1994) Velocity measurements in a high-Reynolds-number momentum-conserving, axisymmetric, turbulent jet. J Fluid Mech 258:31–75
4. Aliabadi A A, Staebler R M, Liu M et al (2016) Characterization and parametrization of Reynolds stress and turbulent heat flux in the stably-stratified lower Arctic troposphere using aircraft measurements. Bound-Lay Meteorol 161:99–126
5. Dally J W, Riley W F, McConnell K G (1993) Instrumentation for engineering measurements. Wiley, New York
6. Aliabadi A A, Moradi M, Clement D et al (2019) Flow and temperature dynamics in an urban canyon under a comprehensive set of wind directions, wind speeds, and thermal stability conditions. Environ Fluid Mech 19:81–109
7. Nambiar M K, Byerlay R A E, Nazem A et al (2020) A Tethered Air Blimp (TAB) for observing the microclimate over a complex terrain. Geosci Instrum Meth Data Syst 9:193–211
8. Martin S, Bange J, Beyrich F (2011) Meteorological profiling of the lower troposphere using the research UAV "M^2AV Carolo". Atmos Meas Tech 4:705–716
9. Palomaki R T, Rose N T, van den Bossche M et al (2017) Wind estimation in the lower atmosphere using multirotor aircraft. J Atmos Ocean Tech 34:1183–1191
10. Thompson N (1980) Tethered Balloons. In: Dobson F, Hasse L, Davis R (ed) Air-sea interaction. Springer, Boston
11. Egerer U, Gottschalk M, Siebert H et al (2019) The new BELUGA setup for collocated turbulence and radiation measurements using a tethered balloon: first applications in the cloudy Arctic boundary layer. Atmos Meas Tech 12:4019–4038
12. Canut G, Couvreux F, Lothon M et al (2016) Turbulence fluxes and variances measured with a sonic anemometer mounted on a tethered balloon. Atmos Meas Tech 9:4375–4386
13. Smith F B (1961) An analysis of vertical wind-fluctuations at heights between 500 and 5,000 ft. Q J Roy Meteor Soc 87:180–193.
14. Thompson N (1972) Turbulence measurements over the sea by a tethered-balloon technique. Q J Roy Meteor Soc 98:745–762
15. Davidson B (1968) The Barbados oceanographic and meteorological experiment. B Am Meteorol Soc 49:928–935

16. Garstang M, La Seur N E (1968) The 1968 Barbados experiment. B Am Meteorol Soc 49:627–635
17. Friedman H A, Callahan W S (1970) The ESSA Research Flight Facility's support of environmental research in 1969. Weatherwise 23:174–185
18. Pollard R T (1978) The Joint Air-Sea Interaction Experiment—JASIN 1978. B Am Meteorol Soc 59:1310–1318
19. Berman E A (1976) Measurements of temperature and downwind spectra in the "Buoyant Subrange". J Atmos Sci 33:495–498
20. Lothon M, Lohou F, Pino D et al (2014) The BLLAST field experiment: Boundary-Layer Late Afternoon and Sunset Turbulence. Atmos Chem Phys 14:10931–10960
21. Byerlay R A E, Nambiar M K, Nazem A et al (2020) Measurement of land surface temperature from oblique angle airborne thermal camera observations. Int J Remote Sens 41:3119–3146
22. Eynard D, Vasseur P, Demonceaux C et al (2012) Real time UAV altitude, attitude and motion estimation from hybrid stereovision. Auton Robot 33:157–172
23. Stull R B (2015) Practical meteorology: An algebra-based survey of atmospheric science. University of British Columbia, Vancouver
24. Ku H H (1966) Notes on the use of propagation of error formulas. J Res Nat Bur Stand Sec C: Eng Inst 70C:263–273
25. Nahian M R, Nazem A, Nambiar M K et al (2020) Complex meteorology over a complex mining facility: Assessment of topography, land use, and grid spacing modifications in WRF. J. Appl Meteorol Clim 59:769–789

Chapter 12
Sonic and Ultrasonic Techniques

Abstract This chapter introduces measurement techniques based on travelling pressure waves in a fluid, namely the sonic and ultrasonic waves. Anemometers based on this concept are described. Theory of speed of sound is established, and the SOnic Detection And Ranging (SODAR) measurement technique is introduced.

12.1 Preliminaries

Sonic and ultrasonic techniques are among the most practical and convenient techniques to measure turbulent flows, particularly those encountered in the atmospheric boundary layer [1]. These techniques rely on the measurement of pressure waves in a fluid system. The main difference between the two approaches is the range of frequencies employed for the measurements. Sonic frequencies are within the human audible range from about 20 Hz to 20 kHz, while the ultrasonic frequencies are above the audible range.

Sonic and ultrasonic techniques for the atmospheric and oceanic measurements have been around since the early 1970s. In these techniques, the periodic cost of maintenance is minimal as the instruments are solid state with no moving parts. The starting threshold and response time, for both flow speed and direction, are essentially zero. These two characteristics make the sonic or ultrasonic techniques good candidates for meteorological, air pollution, and dispersion studies among other applications requiring high accuracy at very low flow speeds [2].

12.2 Sonic and Ultrasonic Anemometers

Improvements in atmospheric measurement technologies have allowed determination of the three-dimensional wind velocity vector with high temporal and spatial resolutions [3]. The very first acoustic anemometers were not actually based on speed of sound but on a wide range of tones that were built in organs to produce sound. The tone then would be varied with wind speed and direction [4]. By the end

© The Author(s), under exclusive license to Springer Nature Switzerland AG 2022
A. A. Aliabadi, *Turbulence*, Mechanical Engineering Series,
https://doi.org/10.1007/978-3-030-95411-6_12

Fig. 12.1 Ultrasonic
anemometers installed on a
30-m mast to measure
turbulence statistics of
atmospheric flow in a rural
area; measurements provide
wind velocity vector
components and sonic
temperature at a frequency of
10 Hz

Fig. 12.2 2D Ultrasonic
anemometers installed on the
mast shown in Fig. 12.1

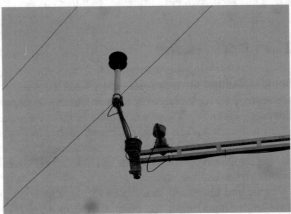

of 1940s the very first sensors were designed to measure temperature with respect to
the propagation of the sound waves [5]. A decade later, [6] explained the theory to
measure the three-dimensional wind velocity vector by sonic or ultrasonic means.

Figure 12.1 shows a modern day tower equipped with ultrasonic anemometers
for atmospheric measurements. The tower is equipped with two 2D ultrasonic
anemometers (upper and lower elevations) detailed in Fig. 12.2 and one 3D ultrasonic anemometer (middle elevation) detailed in Fig. 12.3.

A conventional ultrasonic anemometer generates a small amplitude pressure
disturbance in the fluid, which travels at the speed of a mechanical wave, determined from the physical properties of the fluid. The absolute velocity of pressure
disturbance propagation in a moving fluid is the algebraic sum of the fluid
velocity and the pressure disturbance velocity. Knowing the velocity of the pressure
disturbance, the fluid velocity could then be calculated. Such anemometers need
pairs of acoustic transducers, usually two pairs for two-dimensional velocity vector
measurements and three pairs for three-dimensional velocity vector measurements
[7]. The absolute air velocity along each sonic pathway is calculated by measuring

Fig. 12.3 3D Ultrasonic anemometer installed on the mast shown in Fig. 12.1

the transmission time of a signal along the fixed path. The speed of measurement is determined by the frequency response of the transducer operation [8].

Figure 12.4 shows a schematic of a single-axis anemometer. Suppose that the wind velocity vector is defined by $\mathbf{U} = \mathbf{U}_d + \mathbf{U}_n$ [ms^{-1}], where $\mathbf{U}$ [ms^{-1}] is the horizontal wind velocity vector, $\mathbf{U}_d$ [ms^{-1}] is the wind velocity vector along the transducer path, and $\mathbf{U}_n$ [ms^{-1}] is the wind velocity vector normal to the transducer path. Transducer 1 periodically emits and receives pulses of pressure toward transducer 2. In the case when the magnitude of the wind velocity vector is zero, i.e. $U = 0$ ms^{-1}, the pressure pulse would travel the direct pathway in time $t = d/C$ [s], where d [m] is the separation distance and C [ms^{-1}] is the speed of pressure wave. When there is a component of wind velocity parallel to the path, the pressure pulse travelling time is affected as the pulses are carried along the wind velocity. For instance, if wind blows from transducer 1 to transducer 2, i.e. $\mathbf{U} = \mathbf{U}_d$ [ms^{-1}], then the travelling time would be reduced as the apparent speed of the pressure pulse will be $C + U_d$ [ms^{-1}] while the travelling time from transducer 2 to transducer 1 would be increased as the apparent speed of the pressure pulse will be $C - U_d$ [ms^{-1}]. The wind's normal component has a minor effect on the travelling time. It increases the apparent travel distance or decreases the speed of pressure pulse from C to $C \cos \alpha$ [ms^{-1}], where $\alpha = \arcsin(U_n/C)$ [Rad]. As an example, when $U_n = 20$ ms^{-1}, the apparent speed of pressure pulse would be reduced by 0.17 %. Considering the effects of both parallel and normal components of wind velocity vector, the travel times can be calculated as follows. Taking the difference between the inverse of travel times leads to a relationship between travel times and the component of wind velocity vector along the transducer path. With such a relationship, the component of the wind velocity parallel to the path is determined by

$$t_1 = \frac{d}{C \cos \alpha + U_d}, \tag{12.1}$$

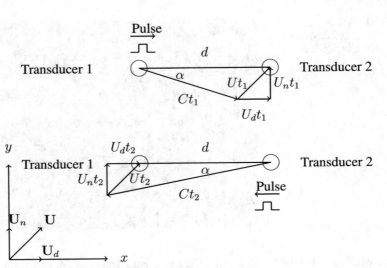

Fig. 12.4 Anemometer vector relation; anemometer measures the time required to transmit a pressure pulse to calculate the wind velocity component along the transducer path; wind velocity vector defined as $\mathbf{U} = \mathbf{U}_d + \mathbf{U}_n$ [ms^{-1}], where $\mathbf{U}$ [ms^{-1}] is the horizontal wind velocity vector, $\mathbf{U}_d$ [ms^{-1}] is the wind velocity vector along the transducer path, and $\mathbf{U}_n$ [ms^{-1}] is the wind velocity vector normal to the transducer path; (top) transducer 1 sends the pulse; (bottom) transducer 2 sends the pulse

$$t_2 = \frac{d}{C \cos \alpha - U_d}, \tag{12.2}$$

$$\frac{1}{t_1} - \frac{1}{t_2} = \frac{C \cos \alpha + U_d}{d} - \frac{C \cos \alpha - U_d}{d}, \tag{12.3}$$

$$U_d = \frac{d}{2} \left(\frac{1}{t_1} - \frac{1}{t_2} \right). \tag{12.4}$$

The robustness of an anemometer's measurement can be verified by calibrating the sensor using wind tunnel testing. Experimental measurements could be conducted comparing one sensor versus another to identify the sensor's measurement error and cross-sensitivities, which are not typically reported by manufacturers [3]. The idea of calibrating an anemometer in a wind tunnel is to find out the cross-instrument difference for each wind vector component as a result of instrument inherent errors and that induced by the mounting fixture. This calibration can be performed along each pathway of a pair of transducers [9].

For atmospheric measurements, the speed of measurement, i.e. sampling frequency, is usually between 10 Hz and 40 Hz. Usually, the higher the sampling rate, the finer time and length scales of turbulence can be detected. Numerous studies report ultrasonic anemometer sampling of the atmosphere at 10 Hz [10–17], 20 Hz

[13, 18–20], or greater than 20 Hz [14, 20–24] for field campaigns focused on turbulence measurements. Ultrasonic anemometers reach accuracies and precisions within centimetres per second.

12.3 SOnic Detection And Ranging (SODAR)

SOnic Detection And Ranging (SODAR) has been used successfully during the last few decades for measurement of atmospheric wind profiles from a few tens of meters up to a few kilometres in altitude. Also known as wind profilers, SODARs are used to measure the scattering of sound waves by atmospheric turbulence. SODAR systems are used to measure wind speed and direction at various heights above the ground and the thermodynamic structure of the lower layer of the atmosphere [25]. A modern day SODAR is shown in Fig. 12.5. SODARs are suited for applications such as wind energy, wind profiling in the airport, and air pollution monitoring [26].

SODARs are equivalent of SOund NAvigation and Ranging (SONAR) systems used in water. As shown in Fig. 12.6, SODARs use the Doppler effect with a multibeam configuration to determine the wind speed and direction. SODARs consist of antennas that transmit and receive acoustic signals. The horizontal components of the wind velocity are calculated from the radially measured Doppler shifts and the specified tilt angle from the vertical direction. The tilt angle, or zenith angle, is generally 15–30°, and the slanted beams have projections on the horizontal plane with right angle (e.g. north and east). Since the Doppler shift of the radial components along the tilted beams includes the influence of both the horizontal and vertical components of the wind, a correction for the vertical velocity is needed in systems with zenith angles less than 20°. In addition, if the system is located in a region where vertical velocities may be greater than about $0.2 \, \text{ms}^{-1}$, corrections for the vertical velocity are needed, regardless of the beam's zenith angle. SODAR

Fig. 12.5 A SODAR installed in a rural area to measure vertical profiles of wind speed from 30 m to 200 m every 30 min

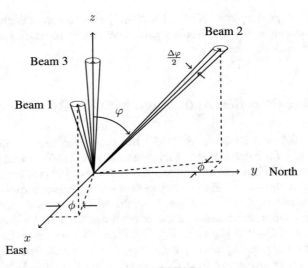

Fig. 12.6 Arrangement of beams in a three-beam SODAR

systems can either use a horn driver and a parabolic dish or an array of speakers to produce a conical beam of sound. A SODAR can have up to five beams with the three-beam configuration as the most common arrangement.

The signal processing unit in SODARs is known as the Acoustic Signal Processor (ASP) having a number of variations including a bank of analog filters with a microprocessor. As the computers are becoming more powerful, additional functions have been assigned to the ASP along with networking, diagnostic responsibilities, satellite communication, and controller features. Moreover, different speaker emitters and detectors are engaged to accomplish a monitoring task that is most often associated with the maximum effective sampling altitude.

In the theory of sound, the velocity of sound [ms^{-1}] in dry (d) and moist (m) air are given by the following equations [27]:

$$C_d = 20.05\sqrt{T} \,, \tag{12.5}$$

$$C_m = C_d \left(1 + 0.14\frac{e}{p} \right) \,, \tag{12.6}$$

where T is the absolute temperature [K] and e/p [–] is the ratio of water vapour pressure e [Pa] to the total pressure p [Pa]. The total contribution of the atmospheric water vapour to the velocity of sound is typically less than 1ms^{-1}, so the effect of humidity perturbations on the sound velocity can be safely ignored in comparison to other effects. The velocity of sound **V** [ms^{-1}] deduced by a stationary observer will be the sum of the velocity of sound **C** [ms^{-1}] relative to air and the velocity of air $\mathbf{U}$ [ms^{-1}]

$$\mathbf{V} = \mathbf{C} + \mathbf{U}.\qquad(12.7)$$

The SODAR works by transmitting sound waves upward into the surroundings and exploiting the Doppler shift in the backscatter signal. The movement of the atmosphere is formed by the wind flow and turbulence that is generated due to mechanical and thermal forces. Thermal turbulence is due to the temperature fluctuations, while mechanical turbulence is due to atmospheric eddies [28]. There are three beams in a SODAR to measure the three components of the wind velocity vector. When the acoustic signal is transmitted from these beams, the sound energy is scattered in all directions by the turbulent refractive index that changes as a function of height. A fraction of the acoustic signal returns toward to the electric transducers as echo. This process provides a continuous-time report of echo strength associated with the turbulence intensity. The height z [m] of the measurement is related to the speed of sound C [ms^{-1}] and transmission time t [s] via

$$z = \frac{Ct}{2},\qquad(12.8)$$

where the constant 2 accounts for the travel of the sound forward and backward from and to the antennas. Note that in a mono-static SODAR, the SODAR system uses the same antenna for transmitting and receiving sound, and the scattering angle between the eddies and the SODAR antenna is 180° [29]. If the scattered sound due to turbulence has a component of motion parallel to the beam direction, this changes the acoustic frequency of the echo (Doppler shift). Analysis of the frequency spectrum of the received backscattered signal allows estimation of the speed of the turbulence parallel with respect to the beam as a function of height. Note that the height can be inferred from the time delay t [s] and knowing the speed of sound C [ms^{-1}].

Some time after the pulse is transmitted, the echo signals will become too weak to be detected by the antennas above the background electrical and acoustic noise, because of spherical spreading of the energy and also atmospheric absorption. At this time, when no longer an echo can be detected, the next acoustic pulse is transmitted, possibly in a different beam direction so that a different component of the wind velocity vector can be estimated. Because of the conical beam shape, the volume occupied by the transmitted acoustic pulse increases as it progresses further away from the SODAR; the echo from height z [m] is received after travelling distance $2z$ [m]; and a scattering layer of finite thickness lengthens the received signal duration, i.e. the time required to receive the echo stretches. For an acoustic pulse duration of τ [s] and beam width $\Delta\varphi$ [Rad], the effective volume $\forall$ [m^3] over which wind speed averaging takes place has a vertical extent of $C\tau/2$ [m] and horizontal radius of $z\Delta\varphi/2$ [m], giving the following expression for the effective volume:

$$\forall = \frac{C\tau}{2}\pi \left(\frac{z\Delta\varphi}{2} \right)^2 . \tag{12.9}$$

It takes some travelling time T [s] to sample the echo signal over a frequency spectrum. This time determines the vertical extent of the effective volume that is being sampled, given by $z = CT/2$ [m]. Note that since the effective volumes for each beam are separated, the components of wind velocity vector are not measured in the same spatial location. Assuming the Kolmogorov spectrum of Eq. 8.13, the scattered acoustic power interpreted by the SODAR is given by [27]:

$$\sigma(\theta, \kappa) = 0.03\kappa^{\frac{1}{3}}\cos^2(\theta) \left[\frac{C_U^2}{C^2}\cos^2\left(\frac{\theta}{2} \right) + 0.13\frac{C_T^2}{T^2} \right] \left[\sin\left(\frac{\theta}{2} \right) \right]^{-\frac{11}{3}} , \tag{12.10}$$

where θ [Rad] is the scatter angle of return ($180°$ or π Rad for mono-static SODARs), T [K] is the ambient temperature, $\kappa = \frac{2\pi}{\lambda}$ [m^{-1}] is the wave number of the acoustic wave (λ [m] is the wave length), and C_T [Km$^{-\frac{1}{3}}$] and C_U [m$^{\frac{2}{3}}$ s^{-1}] are related to the temperature and velocity structure functions, respectively. It must be recalled from Chap. 8 that the structure function is an ensemble average of a signal measured at two different locations in the flow. Resorting to Eq. 8.20, the wind and temperature structure functions are given by [27]:

$$D_{UU} = \langle [U(\mathbf{x}) - U(\mathbf{x} + \Delta\mathbf{x})]^2 \rangle = C_U^2 \Delta\mathbf{x}^{2/3} , \tag{12.11}$$

$$D_{TT} = \langle [T(\mathbf{x}) - T(\mathbf{x} + \Delta\mathbf{x})]^2 \rangle = C_T^2 \Delta\mathbf{x}^{2/3} , \tag{12.12}$$

where $U(\mathbf{x})$ and $U(\mathbf{x} + \Delta\mathbf{x})$ [ms^{-1}] are wind velocity components at points $\mathbf{x}$ and $\mathbf{x} + \Delta\mathbf{x}$ [m], respectively. Likewise $T(\mathbf{x})$ and $T(\mathbf{x} + \Delta\mathbf{x})$ [K] are temperatures at points $\mathbf{x}$ and $\mathbf{x} + \Delta\mathbf{x}$ [m], respectively [25, 27, 30]. These equations allow calculation of the parameters C_T [Km$^{-\frac{1}{3}}$] and C_U [m$^{\frac{2}{3}}$ s^{-1}] once measurements of D_{UU} [m^2 s^{-2}] and D_{TT} [K^2] are available and vice versa. This equation indicates that a full measurement of the scattered power as a function of wave number κ [m^{-1}] and scatter angle θ [Rad] is possible. The scattered acoustic power $\sigma(\theta, \kappa)$ [m^{2}m^{-3}] is per unit volume, per unit incident flux, and per unit solid angle at an angle of θ [Rad] from the initial direction of propagation [27].

The frequency content in the received signal includes electronic and background acoustic noise, Doppler shifted echo signals at frequencies near the transmitted frequency f_T [Hz], and echoes from nearby solid objects such as buildings or masts at the frequency f_T [Hz] (with no Doppler shift). The received signal is demodulated and sampled giving a Doppler Fast Fourier Transform (FFT) spectrum at discrete frequencies. In such a technique the spectrum contains N_f [–] points sampled at a frequency rate of f_s [Hz]. Each spectrum is therefore acquired over a time interval of $P = N_f/f_s$ [s] equivalent to a height interval of $\frac{CN_f}{2f_s}$ [m]. This height resolution for wind velocity vector component estimation is typically around $\Delta z = 10$ m, giving spectral estimates separated by about $\Delta f = 170/\Delta z = 17$ Hz. A transmitted pulse of duration $\tau = 50$ ms gives a spectral peak of half width around 20 Hz, but the returned echo signal contains noise so a peak detection algorithm is required and the estimated Doppler shift frequency is subject to uncertainty. It is common to average the frequency spectrum for each beam over multiple minutes to obtain an improved Signal-to-Noise Ratio (SNR).

With the theory introduced so far, some feasibility calculations are possible. In the case of a mono-static SODAR (backscatter case), Eq. 12.10 can be written as

$$\sigma(\kappa) = 0.0039\kappa^{\frac{1}{3}}\left(\frac{C_T}{T}\right)^2, \tag{12.13}$$

which can be used in a radar equation that relates the received power P_r [W] to the emitting power P [W], scattered acoustic power $\sigma(\kappa)$ [m²m⁻³], speed of sound C [ms⁻¹], sound pulse width τ [s], collecting area of the receiving antenna A_r [m²], range to the scattering region R [m], and attenuation factor L [–], taking into account system inefficiencies. The radar equation can be written as [27]

$$P_r = P\sigma(\kappa)\frac{C\tau}{2}\frac{A_r}{R^2}L. \tag{12.14}$$

Once the wind velocity vector component along each beam is determined, it is possible to use vector calculus to determine the wind velocity vector components with respect to the Cartesian coordinate system shown in Fig. 12.6. Suppose that the wind velocity vector components along the beams are given by U_{B_1}, U_{B_2}, and U_{B_3} [ms⁻¹]. Our goal is to infer the wind velocity components in the Cartesian coordinate system given by $\mathbf{U} = (U, V, W)$ [ms⁻¹]. Unit vectors below specify the direction of each beam:

$$\mathbf{B}_1 = \sin\varphi\cos\phi\mathbf{i} + \sin\varphi\sin\phi\mathbf{j} + \cos\varphi\mathbf{k}, \tag{12.15}$$

$$\mathbf{B}_2 = -\sin\varphi\sin\phi\mathbf{i} + \sin\varphi\cos\phi\mathbf{j} + \cos\varphi\mathbf{k}, \tag{12.16}$$

$$\mathbf{B}_3 = \mathbf{k}. \tag{12.17}$$

Using the dot product operator, it is possible to express the components of $\mathbf{U}$ on each of the directions specified by unit vectors $\mathbf{B}_1$, $\mathbf{B}_2$, and $\mathbf{B}_3$ [–] by the following matrix operation. This operation can also be carried out in reverse, by finding an inverse matrix, to calculate the wind velocity components in the Cartesian coordinate system:

$$\begin{bmatrix} U_{B_1} \\ U_{B_2} \\ U_{B_3} \end{bmatrix} = \begin{bmatrix} \mathbf{U}.\mathbf{B}_1 \\ \mathbf{U}.\mathbf{B}_2 \\ \mathbf{U}.\mathbf{B}_3 \end{bmatrix} = \begin{bmatrix} \sin\varphi\cos\phi & \sin\varphi\sin\phi & \cos\varphi \\ -\sin\varphi\sin\phi & \sin\varphi\cos\phi & \cos\varphi \\ 0 & 0 & 1 \end{bmatrix} \begin{bmatrix} U \\ V \\ W \end{bmatrix}. \tag{12.18}$$

The vertical range of SODARs is approximately 0.2 to 2 km and is a function of frequency, power output, atmospheric stability, turbulence, and, most importantly, the noise environment in which a SODAR is operated. Operating frequencies vary in range from less than 1000 Hz to over 4000 Hz, with power levels up to several hundred watts. SODARs reach accuracies and precisions within centimetres per second. SODARs are most effective in the measurement of mean wind speed and direction. Given the time delay for the acoustic wave to travel back and forth, SODARs are only able to detect large-scale fluctuations of turbulence in the atmosphere.

Figure 12.7 shows the wind speed profiles measured in Turfgrass Institute, Guelph, Canada, in a summer for different classifications of thermal stability (to be defined in Chap. 18) and wind speed. The box plots are created using 5th, 25th, 50th, 75th, and 95th percentiles. It can be seen that the entire measured wind speed profile was correlated with the thermal stability case. Low wind speeds were measured at all altitudes under unstable conditions. This was likely due to the presence of a well-mixed convective surface layer (to be defined in Chap. 18), characterized by a near-constant distribution of wind speed with height due to strong vertical mixing [31]. Under weakly unstable and near neutral cases, high wind speeds were noticed at all altitudes compared to the unstable case. Under the stable case, the wind speed at low altitudes was lower than the weakly unstable and near neutral cases, but at higher altitudes wind speeds as high as the other two cases were noticed. This sharp vertical gradient in the wind speed under thermally stable atmospheric surface layer is typically due to suppressed vertical mixing and has been well documented in the literature [32, 33]. Under very low and low wind conditions, lower wind speeds at low altitudes were observed, while under moderate and high wind conditions, higher wind speeds at low altitudes were observed.

Problems

12.1 An engineer is designing a single-axis ultrasonic anemometer to measure the wind speed by aligning the transducer path along the wind velocity vector. He wants to determine the transducer spacing d [m] such that he can measure the wind speed as fast as $U_d = 5$ ms^{-1} along the path of the transducers with a minimum sampling frequency of $f = 40$ Hz. His signal processing circuits require $t_e = 0.024$ s to record the pair of travelling times measured by the transducers, i.e. a signal consisting of t_1 and t_2 [s]. Assuming that the speed of the pulse in air is $C = 344$ ms^{-1}, what criteria should apply to d [m]? Hint: the normal component of the wind velocity in this setting is zero, so it will have no impact on the travelling time of the pulses, i.e. $\cos \alpha = 1$ [–]. Also consider that the summation of the travel times, t_1, t_2, and the signal processing time t_e [s] should be shorter than the sampling period $T = \frac{1}{f}$ [s], i.e.

$$t_1 + t_2 + t_e = \frac{d}{C + U_d} + \frac{d}{C - U_d} + t_e < \frac{1}{f} . \tag{12.19}$$

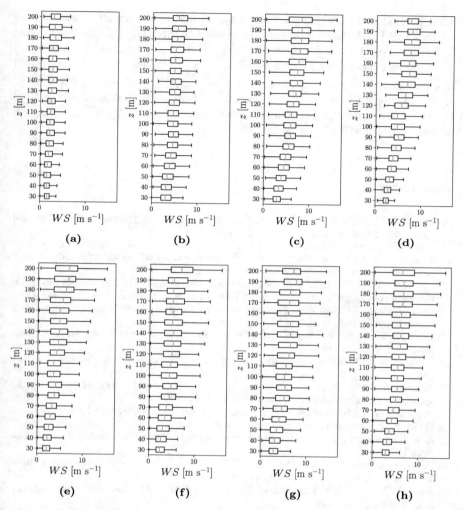

Fig. 12.7 Wind speed profiles measured using a SODAR in Turfgrass Institute, Guelph, Canada, according to thermal stability (**a**, **b**, **c**, and **d**) and wind speed (**e**, **f**, **g**, and **h**) conditions; the box plots are created using 5th, 25th, 50th (orange line), 75th, and 95th percentiles

12.2 A scientist has installed a three-beam SODAR for atmospheric wind profiling. To ease the calculations, she has oriented the SODAR toward the true north such that $\phi = 0°$. In this configuration, beam 1 is toward the east, beam 2 is toward the north, and beam 3 is vertical. The SODAR of interest has the beam zenith angle of $\varphi = 20°$. In matrix form, express the relationship between wind velocity components in the SODAR and Cartesian coordinate systems. Invert the matrix to explicitly express the wind velocity components in the Cartesian coordinate system as a function of the SODAR coordinate system.

12.3   An engineer is designing a SODAR system and needs to calculate the received power P_r [W] in the radar equation [27]. In this SODAR design, the emitting power is $P = 10$ W, the sound pulse width is $\tau = 10^{-2}$ s, the range of scattering is $R = 150$ m, the receiving antenna area is $A_r = 1$ m^2, the wave length of the sound is $\lambda = 2\pi \times 10^{-2}$ m (corresponding to $f = 5$ kHz), $C_T = 4.6 \times 10^{-2}$ Km$^{-\frac{1}{3}}$, ambient temperature is $T = 300$ K, and the system attenuation is $L = 5 \times 10^{-3}$ [–]. Help the engineer calculate the scattered acoustic power $\sigma(\kappa)$ [m^{2}m^{-3}] and the received power P_r [W].

References

1. Kaimal J C, Finnigan J J (1994) Atmospheric boundary layer flows: their structure and measurement. Oxford University Press, Oxford
2. Aliabadi A A, Moradi M, Clement D et al. (2019) Flow and temperature dynamics in an urban canyon under a comprehensive set of wind directions, wind speeds, and thermal stability conditions. Environ Fluid Mech 19:81–109
3. Bärfuss K, Pätzold F, Altstädter B et al (2018) New setup of the UAS ALADINA for measuring boundary layer properties, atmospheric particles and solar radiation. Atmosphere-Basel 9:28
4. Suomi I, Vihma T (2018) Wind gust measurement techniques—From traditional anemometry to new possibilities. Sensors-Basel 18:1300
5. Barrett E W, Suomi V E (1949) Preliminary report on temperature measurement by sonic means. J Meteorol 6:273–276
6. Schotland R M (1955) The measurement of wind velocity by sonic means. J Meteorol 12:386–390
7. Ghaemi-Nasab M, Franchini S, Davari A R et al. (2018) A procedure for calibrating the spinning ultrasonic wind sensors. Measurement 114:365–371
8. Brock F V (2001) Meteorological measurement systems. Oxford University Press, Oxford
9. Grelle A, Lindroth A (1994) Flow distortion by a Solent sonic anemometer: Wind tunnel calibration and its assessment for flux measurements over forest and field. J Atmos Ocean Tech 11:1529–1542
10. Eliasson I, Offerle B, Grimmond C S B et al (2006) Wind fields and turbulence statistics in an urban street canyon. Atmos Environ 40:1–16
11. Klein P, Clark J V (2007) Flow variability in a North American downtown street canyon. J Appl Meteorol Clim 46:851–877
12. Nelson M A, Pardyjak E R, Brown M J et al. (2007) Properties of the wind field within the Oklahoma City Park Avenue street canyon. Part II: spectra, cospectra, and quadrant analyses. J Appl Meteorol Clim 46:2055–2073
13. Ramamurthy P, Pardyjak E R, Klewicki J C (2007) Observations of the effects of atmospheric stability on turbulence statistics deep within an urban street canyon. J Appl Meteorol Clim 46:2074–2085
14. Nelson M A, Pardyjak E R, Klewicki J C et al. (2007) Properties of the wind field within the Oklahoma City Park Avenue street canyon. Part I: mean flow and turbulence statistics. J Appl Meteorol Clim 46:2038–2054
15. Balogun A A, Tomlin A S, Wood C R et al. (2010) In-street wind direction variability in the vicinity of a busy intersection in central London. Bound-Lay Meteorol 136:489–513
16. Zajic D, Fernando H J S, Calhoun R et al. (2011) Flow and turbulence in an urban canyon. J Appl Meteorol Clim 50:203–223
17. Klein P M, Galvez J M (2015) Flow and turbulence characteristics in a suburban street canyon. Environ Fluid Mech 15:419–438

18. Barlow J F, Halios C H, Lane S E et al. (2015) Observations of urban boundary layer structure during a strong urban heat island event. Environ Fluid Mech 15:373–398
19. Blackman K, Perret L, Savory E et al. (2015) Field and wind tunnel modeling of an idealized street canyon flow. Atmos Environ 106:139–153
20. Giometto M G, Christen A, Meneveau C et al. (2016) Spatial characteristics of roughness sublayer mean flow and turbulence over a realistic urban surface. Bound-Lay Meteorol 160:425–452
21. Louka P, Belcher S E, Harrison R G (2000) Coupling between air flow in streets and the well-developed boundary layer aloft. Atmos Environ 34:2613–2621
22. Rotach M W, Vogt R, Bernhofer C et al. (2005) BUBBLE–an urban boundary layer meteorology project. Theor Appl Climatol 81:231–261
23. Inagaki A, Kanda M (2008) Turbulent flow similarity over an array of cubes in near-neutrally stratified atmospheric flow. J Fluid Mech 615:101–120
24. Inagaki A, Kanda M (2010) Organized structure of active turbulence over an array of cubes within the logarithmic layer of atmospheric flow. Bound-Lay Meteorol 135:209–228
25. Beyrich F (1997) Mixing height estimation from SODAR data - a critical discussion. Atmos Environ 31:3941–3953
26. Chan P W (2008) Measurement of turbulence intensity profile by a mini-SODAR. Meteorol Appl 15:249–258
27. Little C G (1969) Acoustic methods for the remote probing of the lower atmosphere. Proceedings of the IEEE 57:571–578
28. Garratt J (1994) The Atmospheric Boundary Layer. Cambridge University Press, Cambridge
29. Behrens P, O'Sullivan J, Archer R et al. (2012) Underestimation of monostatic SODAR measurements in complex terrain. Bound-Lay Meteorol 143:97–106
30. Keder J, Foken T, Gerstmann W et al. (1989) Measurement of wind parameters and heat flux with the sensitron Doppler SODAR. Bound-Lay Meteorol 46:195–204
31. Kaimal J C, Wyngaard J C, Haugen D A et al. (1976) Turbulence structure in the convective boundary layer. J Atmos Sci 33:2152–2169
32. Nieuwstadt F T M (1984) The turbulent structure of the stable, nocturnal boundary layer. J Atmos Sci 41:2202–2216
33. Mahrt L, Vickers D (2006) Extremely weak mixing in stable conditions. Bound-Lay Meteorol 119:19–39

Chapter 13
Electro-magnetic Techniques

Abstract This chapter introduces the measurement techniques that rely on travelling electro-magnetic waves. A handful of techniques are described that require multiphase systems comprised of fluids and suspended particles (or bubbles), involving shadowgraphy, Particle Tracking Velocimetry (PTV), and Particle Image Velocimetry (PIV). The Schlieren imaging is described that is suitable for measuring flows exhibiting density variations. The Laser Doppler Velocimetry (LDV) technique is introduced, which relies on the Doppler shift in a laser beam to measure velocity in transparent or semi-transparent fluid flows. Radiometry and pyrometry techniques are discussed to measure surface temperature, which rely on radiative exchange between the surface and a sensor. Finally, the Light Detection And Ranging (LiDAR) technique is introduced, which uses the Doppler shift principle over long distances to measure flow properties such as the velocity vector components.

13.1 Overview

Electro-magnetic techniques in measuring turbulent flows all have one characteristic in common: they all function on principles of electro-magnetic radiation. The electro-magnetic source and detection technology in each individual technique may vary.

13.2 Shadowgraphy

Shadowgraphy is among the most basic forms of flow measurement techniques. The underlying principle in shadowgraphy is that light rays always travel in a straight line; however, their path will be altered in non-homogeneous fluids. For instance, obstacles and density variations in the fluid cause light to reflect or refract, hence form a shadow of some sort if projected on a screen [1].

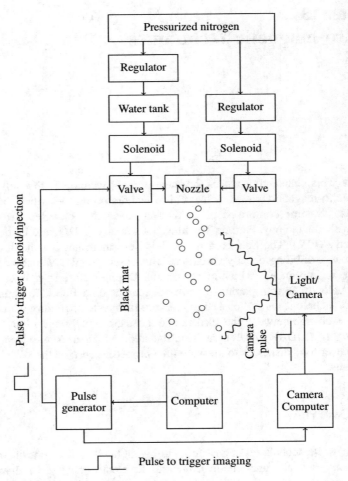

Fig. 13.1 Schematic of a shadowgraphy experiment [2]

Figure 13.1 shows the schematic of a shadowgraphy experiment to measure the axial and radial penetration of a transient spray generated by a nozzle [2]. A *back lighting* arrangement was used to take images of the spray using a camera. A pulsed laser (532 nm) was used as a light source to shine a collimated beam of green light. The beam was passed through a lens so that the beam was traced as a round circle on a container of diffuse and fluorescent medium (liquid Rhodamine). This medium further expanded the beam to cover the test area. Due to the presence of liquid droplets, a portion of the incident light was reflected and measured by the camera, while the remainder of the incident light did not reflect from the background black mat. This enabled detecting the spray in the image. The firing of the laser pulse and the camera imaging were controlled by trigger pulses from a Programmable

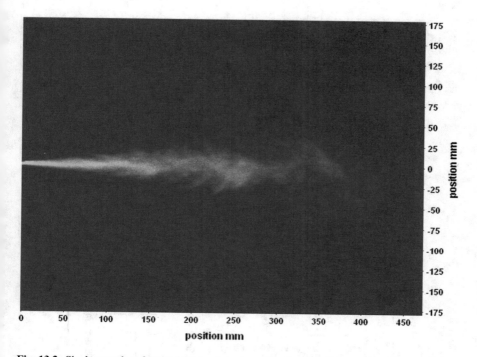

position mm

position mm

Fig. 13.2 Shadowgraphy of a transient droplet spray using the experiment outline in Fig. 13.1 [2]

Timing Unit (PTU), which actuated the valves for the nozzle operation as well as the imaging with a time delay.

Figure 13.2 shows the shadowgraphy of a transient droplet spray, using the experimental setup shown in Fig. 13.1. Such images may be repeated with different time delays with respect to the nozzle valve actuation time to provide the transient spray axial and redial velocity components on the leading and trailing edges of the spray.

13.3 Particle Tracking Velocimetry (PTV)

In Particle Tracking Velocimetry (PTV), the flow is seeded with small particles, which faithfully follow the fluid, while individual particles are tracked to produce the velocity field and other relevant flow statistics. In the simplest form, PTV acquires two consecutive images (with a very small time delay) of flow field seeded by these tracer particles, and the particle images are then cross-correlated to track movement of individual particles and hence determination of the flow field.

Figure 13.3 shows the schematic of a PTV experiment used to measure flow velocity of a transient spray from a nozzle [2]. In this setup a dual-head laser

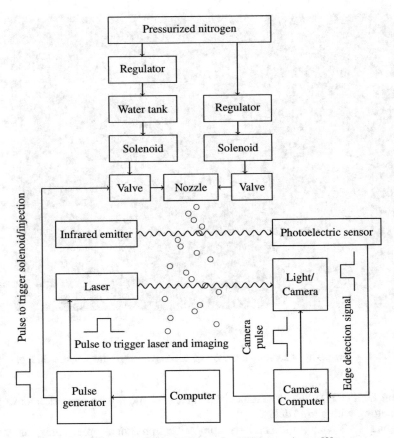

Fig. 13.3 Schematic of a Particle Tracking Velocimetry (PTV) experiment [2]

(532 nm) generated two light pulses separated by 10 μs. A camera was used to grab a double-frame image timed with the laser firing. The two successive images taken were then used to track individual particles for velocity calculation. To obtain a precise timing for illumination and imaging of the spray, a pair of photoelectric sensor and an infrared emitter were arranged in a *through-beam* setup at the exit of the spray near the nozzle tip. If the spray blocked the infrared beam path, the receiver would produce a triggering signal for the PTU board, which was programmed to trigger the laser and camera for imaging given specific time delays. Figures 13.4 and 13.5 show the images of particles and their identification for the PTV experiment shown in Fig. 13.3. The particles are identified, with their diameters in [mm], and tracked for spray velocity calculation.

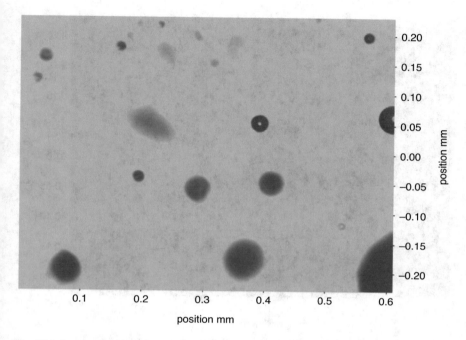

Fig. 13.4 Image of particles in a Particle Tracking Velocimetry (PTV) experiment

13.4 Particle Image Velocimetry (PIV)

Particle image velocimetry (PIV) is an optical method of flow visualization where one can obtain instantaneous velocity measurements and related properties in fluids. The fluid is seeded with tracer particles, which, for sufficiently small particles, are assumed to faithfully follow the flow dynamics (the degree to which the particles faithfully follow the flow is represented by the Stokes number). The fluid with entrained particles is illuminated so that particles are visible. The motion of the seeding particles is used to calculate speed and direction (the velocity field) of the flow being studied [3, 4].

Other similar techniques used to measure flows are Laser Doppler Velocimetry (LDV) and Particle Tracking Velocimetry (PTV). The main difference between PIV and those techniques is that PIV produces two-dimensional or even three-dimensional vector fields, while the other techniques measure the velocity at a point. For PIV, the particle concentration is such that it is possible to identify individual particles in an image, but not with certainty to track them between images. In its simplest form, PIV acquires two consecutive images (with a very small time delay) of flow field seeded by these tracer particles, and the particle images are then cross-correlated to yield the instantaneous fluid velocity field.

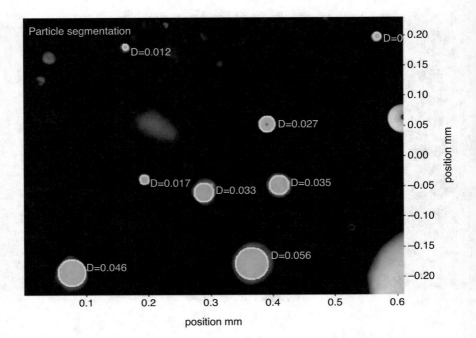

Fig. 13.5 Segmentation of the particles shown in Fig. 13.4; particle diameters D [mm] are identified in the PTV experiment

A typical PIV apparatus consists of a camera (normally a digital camera with a CCD chip in modern systems), a strobe or laser with an optical arrangement to limit the physical region illuminated (normally a cylindrical lens to convert a light beam to a sheet), a synchronizer to act as an external trigger for control of the camera and laser, the seeding particles, and the fluid under investigation. Normally a PIV software is used to post-process the optical images to obtain the velocity field [2].

The camera is normally placed facing the light sheet generated by a laser. As seed particles pass through the laser light sheet, they scatter light, which is captured by the camera, so the particles can be detected. The laser timing is controlled using a synchronization unit to ensure each laser pulse occurs during the exposure time of each frame captured by the camera. By varying the time between consecutive laser pulses (Δt [s]) and the camera's frame rate, a wide range of the scales of fluid motion can be detected [4, 5].

PIV analysis is performed on pairs of images. The primary component of PIV image analysis is the cross-correlation. The first image in a given pair is split into many *interrogation windows*, while the second image, taken with a Δt [s] time delay, is split into many larger *search windows*. Each interrogation window corresponds to a unique search window. A sample image pair with one interrogation window and one search window is shown in Fig. 13.6. The different sizes of the interrogation and search windows enable the interrogation window to rigidly

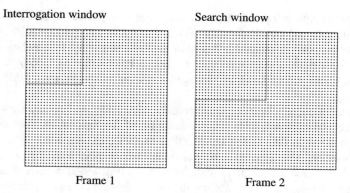

Fig. 13.6 Interrogation and search windows in a PIV system; particles are shown using dots; actual particle distribution in a PIV image is not uniform

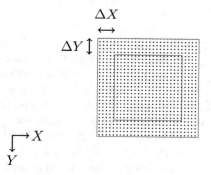

Fig. 13.7 Translation of interrogation window within the corresponding search window in a PIV system

translate within its corresponding search window as shown in Fig. 13.7. The two-dimensional cross-correlation between the interrogation and translated interrogation windows is calculated for all possible translations $(\Delta X, \Delta Y)$ [(m, m)] within the search window. The highest correlation provides the best likelihood for the translation of particles in the original interrogation window. The location for this interrogation window can be given by $(\Delta X_{max}, \Delta Y_{max})$ [(m, m)]. Subsequently, the velocity components in the two-dimensional plane can be given by [4, 5]:

$$(U, V) = \left(\frac{\Delta X_{max}}{\Delta t}, \frac{\Delta Y_{max}}{\Delta t} \right) . \tag{13.1}$$

To construct the actual velocity field, it is necessary to try many pairs of interrogation and search windows. In fact it is possible to slide both windows in the X and Y directions to cover the entire frame. One limitation of planar PIV is the ability to measure only two components of velocity. While the laser sheet can be moved to provide velocity fields over several planes, the out-of-plane velocity can never be measured simultaneously with the other two components. A solution to this issue has been developed in the form of the Stereoscopic PIV technique. In this technique the three-dimensional velocity field is constructed by having two off-axis cameras, which record particles' movement within a laser sheet of finite thickness [5].

Physics of light scattering is at the core of the PIV technique. Generally, the light scattered by small particles is a function of the ratio of the refractive index of the particles to that of the surrounding medium, the particle's size, its shape, and orientation. The light scattering also depends on polarization and observation angle [6]. For spherical particles with diameter d_p [m], which is larger than the laser wavelength λ [m], the Mie scattering theory can describe the physical process. The Mie scattering theory characterizes the normalized diameter of the particle as [6]

$$q = \frac{\pi d_p}{\lambda} ,$$

(13.2)

where if q [–] is larger than unity, approximately q [–] local maxima appear for scattered light intensity over the angular distribution around the particle from 0 to 180°. With increasing q [–], the ratio of forward to backward scattering intensity will increase, and so the angular position of the camera with respect to the light source and laser sheet can be decided accordingly for optimal imaging. Also, with increasing particle diameter the intensity of the scattered light will increase. The average intensity of scattered light roughly increases with q^2 [–] [6].

Common particles used for PIV in liquid flows are made of polystyrene, aluminium flakes, hollow glass spheres, and different oils, with diameters in the range $d_p \sim 2 - 5000\,\mu\text{m}$. For gas flows, common particles are made of polystyrene, alumina, titania, smoke, and different oils, with diameters in the range $d_p \sim 0.1 - 10\,\mu\text{m}$. In some cases hollow particles are used, either for liquid or gas flow analysis, which are made of hollow glass spheres and oxygen bubbles. In such cases much larger particle diameters are possible [6].

The first laser was built in 1960 by Theodore Harold Maiman (1927–2007) at Hughes Research Laboratories. A laser is a device that emits light through a process of optical amplification based on the stimulated emission of electro-magnetic radiation. The word *laser* is an acronym for *light amplification by stimulated emission of radiation*. Laser beams are spatially and temporally coherent. The light intensity in many laser beams is distributed roughly according to the Gaussian shape, that is the centre of the beam is at the highest intensity, while the intensity will decrease away from the centre in the radial direction. Table 13.1 shows various laser types used in PIV. PIV lasers range from very low power (e.g. semiconductor lasers) to those of very high power (e.g. Argon-ion lasers). According to the light intensity and ability to control the light emission pulse rate, lasers can be used for low (e.g. Helium-neon lasers) or high (e.g. Neodym-YLF lasers) speed imaging analysis [6].

Many optical arrangements are possible to create a light sheet for PIV. The essential component for creation of the light sheet is a cylindrical lens. When using

Table 13.1 Laser types commonly used in PIV [6]

Name	Symbol	Wavelength [nm]
Helium-neon lasers	He-Ne	$\lambda = 633$
Copper-vapour lasers	Cu	$\lambda = 510, 578$
Argon-ion lasers	Ar^+	$\lambda = 514, 488$
Ruby lasers	Cr^{3+}	$\lambda = 694$
Neodym-YAG lasers	Nd:YAG	$\lambda = 1064, 532$
Neodym-YLF lasers	Nd:YLF	$\lambda = 1053, 526$

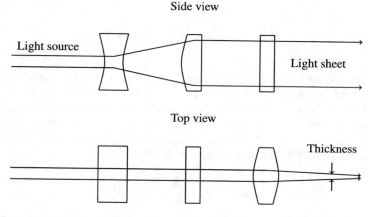

Fig. 13.8 Light sheet optics based on three cylindrical lenses (one with a negative focal length) used in PIV [6]

lasers with a very fine diameter, only one cylindrical lens may be sufficient to generate the light sheet of desired thickness. However, many light sources (e.g. Nd:YAG lasers) may have a wide beam of a coarse diameter. In this case more lenses are needed to diverge the beam on the one hand, and, at the same time, to focus it for generation of a light sheet of desired thickness on the other hand. Figure 13.8 shows such an arrangement. For high power lasers, it is recommended to diverge the beam at the closest distance to the laser source, such that air ionization can be avoided and safety risks can be minimized [6].

The mathematics of PIV using two singly exposed recordings can be elucidated in a simplified manner [6]. Consider the two-dimensional PIV setup shown in Fig. 13.7, where the light sheet is very thin so that particle velocity components in two directions of X and Y can be measured. Suppose that $\mathbf{\Gamma} = (\mathbf{X}_1, \mathbf{X}_2, \ldots, \mathbf{X}_N)$ holds the locations of N tracer particles in the interrogation window, where $\mathbf{X}_i = (X_i, Y_i)$ [(m, m)] are coordinates of each particle. The coordinates in the image may be scaled by a factor M [–], such that the image coordinates for each particle can be shown using $\mathbf{x}_i = (x_i, y_i) = (MX_i, MY_i)$ [(m, m)]. Let us assume a constant displacement $\mathbf{D} = (D_x, D_y)$ [(m, m)] of all particles inside the interrogation window, so that the particle locations during the second exposure at time $t' = t + \Delta t$

[s] can be given by:

$$\mathbf{X}'_i = \mathbf{X}_i + \mathbf{D} , \tag{13.3}$$

where $\mathbf{X}_i = (X_i, Y_i)$ [(m, m)] is the position of the particle i in the first exposure, and $\mathbf{X}'_i = (X_i + D_x, Y_i + D_y)$ [(m, m)]. The displacement in the image $\mathbf{d} = (d_x, d_y)$ [(m, m)] can also be assumed to scale uniformly with actual displacement, such that $\mathbf{d} = (d_x, d_y) = (M D_x, M D_y)$ [(m, m)], where M [–] is the scaling factor. The image normalized intensity field of the first exposure can be written as

$$I(\mathbf{x}, \mathbf{\Gamma}) = \sum_{i=1}^{N} V_0(\mathbf{X}_i) \tau(\mathbf{x} - \mathbf{x}_i) , \tag{13.4}$$

where we note that $V_0(\mathbf{X}_i)$ [–] is a transfer function giving the normalized light energy of the image at individual particle position of $\mathbf{X}_i$ [(m, m)]. This transfer function should be considered because light intensity may change throughout the image based on the optical design of the experiment. We also note that $\tau(\mathbf{x} - \mathbf{x}_i)$ [–] is a point spread function of normalized light intensity for each particle positioned at $\mathbf{x_i}$ [(m, m)], which is usually Gaussian in shape

$$\tau(\mathbf{x} - \mathbf{x}_i) = K \exp\left(-\frac{8|\mathbf{x} - \mathbf{x}_i|^2}{d_\tau^2}\right) , \tag{13.5}$$

where K [–] and d_τ [m] depend on the experimental setup. This equation expresses the idea that the light intensity scattered by a particle drops according to the Gaussian formula as a function of distance away from the centre of the particle. The image normalized intensity field of the second exposure can be written as

$$I'(\mathbf{x}, \mathbf{\Gamma}) = \sum_{j=1}^{N} V_0(\mathbf{X}_j + \mathbf{D}) \tau(\mathbf{x} - \mathbf{x}_j - \mathbf{d}) , \tag{13.6}$$

where $V_0(\mathbf{X}_j + \mathbf{D})$ [–] defines the interrogation volume for the second exposure, which may be different than that of the first exposure. The cross-correlation function of the two exposures can be written as

$$R_{II}(\mathbf{s}, \mathbf{\Gamma}, \mathbf{D}) = \frac{1}{a_I} \sum_{i,j} V_0(\mathbf{X}_i) V_0(\mathbf{X}_j + \mathbf{D}) \int_{a_I} \tau(\mathbf{x} - \mathbf{x}_i) \tau(\mathbf{x} - \mathbf{x}_j + \mathbf{s} - \mathbf{d}) d\mathbf{x}$$

$$= \sum_{i,j} V_0(\mathbf{X}_i) V_0(\mathbf{X}_j + \mathbf{D}) R_\tau(\mathbf{x}_i - \mathbf{x}_j + \mathbf{s} - \mathbf{d}) , \tag{13.7}$$

where $\mathbf{s}$ [(m, m)] is the separation vector between the two windows. The two-dimensional integral is taken over the spatial domain of the windows and written using the function $R_\tau(\mathbf{x}_i - \mathbf{x}_j + \mathbf{s} - \mathbf{d})$ [m^2]. At this point we need to distinguish the $i \neq j$ terms, which represent the correlation of different randomly distributed particles and therefore mainly noise in the correlation plane, from the $i = j$ terms, which contain the displacement information desired. We can therefore write

$$R_{II}(\mathbf{s}, \boldsymbol{\Gamma}, \mathbf{D}) = \underbrace{\sum_{i \neq j} V_0(\mathbf{X}_i) V_0(\mathbf{X}_j + \mathbf{D}) R_\tau(\mathbf{x}_i - \mathbf{x}_j + \mathbf{s} - \mathbf{d})}_{R_C(\mathbf{s}, \boldsymbol{\Gamma}, \mathbf{D}) + R_F(\mathbf{s}, \boldsymbol{\Gamma}, \mathbf{D})}$$

$$+ \underbrace{R_\tau(\mathbf{s} - \mathbf{d}) \sum_{i=1}^{N} V_0(\mathbf{X}_i) V_0(\mathbf{X}_i + \mathbf{D})}_{R_D(\mathbf{s}, \boldsymbol{\Gamma}, \mathbf{D})}, \tag{13.8}$$

which has been written as the sum of three parts, where $R_D(\mathbf{s}, \boldsymbol{\Gamma}, \mathbf{D})$ [m^2] represents the component of the cross-correlation function that corresponds to the correlation of images of particles obtained from the first exposure with images of identical particles obtained from the second exposure ($i = j$ terms). Therefore, for a given distribution of particles inside the flow, the displacement correlation peak reaches a maximum for $\mathbf{s} = \mathbf{d}$ [m, m]. Knowing this $\mathbf{s} = (s_x, s_y)$ [m, m], we can write

$$(\Delta X_{max}, \Delta Y_{max}) = \left(\frac{s_x}{M}, \frac{s_y}{M}\right), \tag{13.9}$$

which would ultimately allow calculation of velocity components according to Eq. 13.1. The PIV technique has been deployed on large scales beyond laboratory experiments using the naturally occurring particles. For instance snowflakes have been used to detect atmospheric flows around large-scale wind turbines [7].

13.5 Schlieren Imaging

Unlike PIV and PTV, Schlieren imaging is truly a nonintrusive technique that relies on the fact that the change in refractive index causes light to deviate due to optical inhomogeneities present in the medium. Schlieren methods can be used for a broad range of high-speed turbulent flows containing refractive index gradients in the form of identifiable and distinguishable flow structures. In Schlieren imaging, the eddies in a turbulent flow field serve as PIV particles. Unlike PIV, there are no seeding particles in Schlieren imaging. As the eddy length scale decreases with the increasing Reynolds number, the length scales of the turbulent eddies become exceptionally important. These self-seeded successive Schlieren images with a small time delay between them can be correlated to find velocity field information. Thus, the analysis of Schlieren images is of great importance in the field of fluid mechanics since this system enables the visualization and flow field calculation of unseeded flow [8].

In classical Schlieren photography, the collimated light is focused with a converging optical element (usually a lens or curved mirror), and a knife-edge is placed at the focal point, positioned to block about half the light. In flow of uniform density this will simply make the photograph half as bright. However, in flow with density variations the distorted beam focuses imperfectly, and parts that have been focused in an area covered by the knife-edge are blocked. The result is a set of lighter and darker patches corresponding to positive and negative fluid density gradients in the direction normal to the knife-edge. When a knife-edge is used, the system is generally referred to as a Schlieren system, which measures the first derivative of density in the direction of the knife-edge. If a knife-edge is not used, the system is generally referred to as a shadowgraph system, which measures the second derivative of density.

If the fluid flow is uniform, the image will be steady, but any turbulence will cause scintillation, the shimmering effect that can be seen on hot surfaces on a sunny day. To visualize instantaneous density profiles, a short-duration flash (rather than continuous illumination) may be used.

For air and other gases there is a linear relationship between the refractive index n [–] and the gas density ρ [gcm^{-3}] via

$$n - 1 = k\rho , \tag{13.10}$$

where k [cm^{3}g^{-1}] is the Gladstone–Dale coefficient . For air $k = 0.23$ cm^{3}g^{-1} under standard conditions. n [–] is only very weakly dependent on ρ [gcm^{-3}]. A change in air density of two orders of magnitude causes only about 3% change in n [–]. Therefore, to detect small density differences in a gas very sensitive optical experiments are needed. The refractivity of a gas is $n - 1$ [–], and it depends on gas composition, temperature, density, and the wavelength of the illumination. For many cases the gas may be modelled as an ideal gas where $P/\rho = RT$ [Jmol^{-1}], in which P [Pa] is pressure, R [JK^{-1}mol^{-1}] is the specific gas constant, T [K] is absolute temperature, and ρ [molm^{-3}] is molar density. k [cm^{3}g^{-1}] also increases slightly with increasing light wavelength, so using light sources with wavelength closer to the infrared range enables enhanced optical sensitivity [1].

Our interest is to find the bending or refraction of light rays as they pass through non-homogeneous media. Figure 13.9 shows the refraction of light through the non-homogeneous media, in which density varies along the x and y directions in a region of space. Here the light ray of interest is that which travels in the z direction, i.e. the main optical pathway for light. It can be shown that the optical inhomogeneities bend light rays in proportion to their gradients of refractive index in the $x - y$ plane. The resulting ray curvatures can be given by [1]:

$$\frac{\partial^2 x}{\partial z^2} = \frac{1}{n}\frac{\partial n}{\partial x} \text{ and } \frac{\partial^2 y}{\partial z^2} = \frac{1}{n}\frac{\partial n}{\partial y} . \tag{13.11}$$

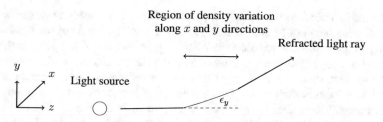

Fig. 13.9 Refraction of light as it passes through non-homogeneous media

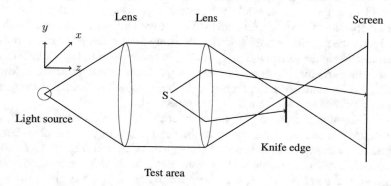

Fig. 13.10 Diagram of a simple Schlieren system with a point light source

Integrating once, the components of the angular ray deflection in the x and y directions can be given by:

$$\epsilon_x = \frac{1}{n} \int \frac{\partial n}{\partial x} dz \text{ and } \epsilon_y = \frac{1}{n} \int \frac{\partial n}{\partial y} dz \,, \tag{13.12}$$

where the integration is made along the z direction over the region of space in which density variations exist. For two-dimensional Schlieren of extent L [m] along the optical axis, the deflection becomes

$$\epsilon_x = \frac{L}{n_0} \frac{\overline{\partial n}}{\partial x} \text{ and } \epsilon_y = \frac{L}{n_0} \frac{\overline{\partial n}}{\partial y} \,, \tag{13.13}$$

where the over bars now indicate the average gradient of the refractive index in the region of space where density gradients exist. Here n_0 [–] is the refractive index of the surrounding medium.

We can now describe the simple lens-type Schlieren system using a point light source. Figure 13.10 shows the setup of the system. Using the first lens, the rays from point light source are collimated. A second lens focuses the beam to an image of the point source. The beam is further projected on the screen for viewing, where a real inverted image would form. A knife-edge is positioned close to the focus of the second lens before the screen. As the knife-edge advances towards the focal point, nothing happens until it rather suddenly blocks the image of the light source, causing a dark screen. In practice, the knife-edge is positioned just prior to the focal point, so an image is obtained on the screen. If a Schlieren object S is positioned in the test area, it would bend the light rays from the original path. Nevertheless, the second lens focuses the ray from each point on S to a corresponding point on the screen image. Two example rays from S are shown in the figure. The upward ray brightens the point on the screen, but the downward ray is blocked by the knife-edge. A finite Schlieren object refracts many such rays in many directions. All downward rays are blocked, while the upward rays form a partial picture of the Schlieren object on the screen. The orientation of the knife-edge matters. As shown in the figure, the positioning of the knife-edge only detects the vertical components of the refractive index gradient, i.e. $\frac{\partial n}{\partial y}$ [m^{-1}], in accordance to Fig. 13.9. With this setup, a Schlieren object with only horizontal gradients of the refractive index, i.e. $\frac{\partial n}{\partial x}$ [m^{-1}], will remain invisible. This may justify forming two Schlieren images for the better characterization of the Schlieren object [1].

The point source Schlieren is not very practical. Its operation is on-or-off, with on corresponding to a light path close to the central axis of the optical system. To take real measurements, an extended light source is needed. Figure 13.11 shows the configuration for an extended light source system. In this set up an extra lens is added after the knife-edge. Again, the principal optical path is along the z axis and the x and y axes point to the transverse directions. The extended light source is shown with a downward arrow although in practice it can be an illuminated area. Collimated by the first lens, the light beam travels through the test section and is refocused using the second lens, so that it forms an inverted image at the knife-edge. Note that since the light source is no longer a point, collimation no longer produces exactly parallel rays. In fact, an array of point sources can be imagined that are distributed over the vertical y axis. Each point source forms a Schlieren beam that focuses on a corresponding point in the light source image at the knife-

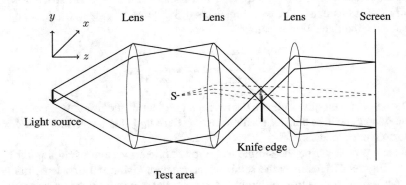

Fig. 13.11 Diagram of a simple Schlieren system with an extended light source

edge. Here the extreme rays from the top and bottom of the light source are traced in the figure.

The knife-edge cuts off a portion of the image of the extended light source. Beyond the knife-edge, a third lens focuses an inverted image of the Schlieren test area on a viewing screen. The light source and knife-edge planes, as well as the test area and screen planes, form sets of conjugate optical planes. These pairs of conjugate planes are important in understanding the Schlieren effect, for what appears in one also appears undistorted in its mate, but for a possible scale transformation. Thus an accurate image of the light source is focused on the knife-edge, and a true image of the Schlieren object S appears on the viewing screen.

The light source now has a finite size, and every point in the test area is illuminated by numerous rays within a cone limited by the extreme rays of the source. In fact, each point in the light source illuminates many test area points. This provides depth of focus to the Schlieren system, which was lacking in the point source system.

Now consider one test area to be subject to refraction by the Schlieren object S. Whereas in the single point source system only one light ray was bent, now a ray bundle from all point sources passing through the area will be bent. This bundle is shown using dashed lines in the figure. All these rays refract by an angle close to ϵ_y [m m^{-1}]. As usual, the downward portion of the ray bundle is blocked by the knife-edge, while the upward portion of the ray bundle continues to pass the third lens and reaches the screen. The Schlieren image of the collection of objects S will be built on the screen from many bundles of light rays that result in enhanced or suppressed illuminance. This process enables a continuous grey-scale Schlieren image, rather than merely a binary black-and-white image.

Note that a bare bulb source is seldom adequate for Schlieren imaging. Instead, it is common practice to use a condenser lens and slit to define a bright effective source with sharp boundaries at the front focus of the first Schlieren lens. Finally, the lenses and the knife-edge can be positioned in such a way to control the scale of the magnification of the image from the test section.

13.6 Laser Doppler Velocimetry (LDV)

Laser Doppler Velocimetry (LDV), also known as Laser Doppler Anemometry (LDA), is the technique of using the Doppler shift in a laser beam to measure the velocity in transparent or semi-transparent fluid flows. The measurement with LDV or LDA is absolute, linear with velocity, and requires no pre-calibration.

In its simplest form, LDV crosses two beams of collimated, monochromatic, and coherent laser light in the flow of the fluid being measured. The two beams are usually obtained by splitting a single beam, thus ensuring coherence between the two. Lasers with wavelengths in the visible spectrum (390–750 nm) are commonly used; these are typically He-Ne, Argon ion, or laser diode, allowing the beam path to be observed. A transmitting lens focuses the beams to intersect at their waists

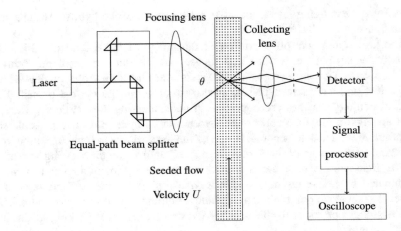

Fig. 13.12 Schematic of a Laser Doppler Velocimetry (LDV) experiment

(the focal point of a laser beam), where they interfere and generate a set of straight fringes. As particles (either naturally occurring or induced) entrained in the fluid pass through the fringes, they reflect light that is then collected by a receiving optic and focused on a photodetector (typically an avalanche photodiode).

The reflected light fluctuates in intensity, the frequency of which is equivalent to the Doppler shift between the incident and scattered light, and is thus proportional to the component of particle velocity, which lies in the plane of two laser beams. If the sensor is aligned to the flow such that the fringes are perpendicular to the flow direction, the electrical signal from the photodetector will then be proportional to the full particle velocity. By combining three devices (i.e. employing three lasers: He-Ne, Argon ion, and laser diode) with different wavelengths, all three flow velocity components can be simultaneously measured [9].

Figure 13.12 shows the schematic of an LDV experiment. The focusing lens focuses the laser beam at a point where velocity is to be measured. Sensing with a photodetector the light scattered by tiny particles, carried along with the fluid as it passes through the laser focal point, is measured. The velocity of the particles causes a Doppler shift of the scattered light frequency. The Doppler shift frequency is directly related to the particle velocities. The velocity U [ms^{-1}] can be given by:

$$U = \frac{\lambda}{2\sin(\theta/2)} f_D ,$$
(13.14)

where λ [m] is the wavelength of the laser beam, θ [Rad] is the angle between the two intersecting light beams from the equal-path beam splitter, and f_D [Hz] is the Doppler shift frequency. Using LDV, the measurement of velocity is direct and not influenced by heat transfer or pressure of the fluid. No physical objects need to be inserted in the flow, so the flow is undisturbed. LDV senses the velocity in a very small volume of the fluid, so the spatial resolution can be very high. However, an

LDV experiment needs a transparent flow channel, requires tiny seed particles in the flow, and can be very expensive.

13.7 Radiometry and Pyrometry

Radiometry is a set of techniques for measuring electro-magnetic radiation, including visible light. Radiometric techniques in optics characterize the distribution of the radiation's power in space, as opposed to photometric techniques, which characterize the light's interaction with the human eye. Radiometry is distinct from quantum techniques such as photon counting. The use of radiometers to determine the temperature of objects and gasses by measuring radiation flux is called Pyrometry. Handheld pyrometer devices are often marketed as infrared thermometers. Radiometry is important in astronomy, especially radio astronomy, and plays a significant role in earth remote sensing. The measurement techniques categorized as radiometry in optics are called photometry in some astronomical applications, contrary to the optics usage of the term.

Planck's radiation law was a mathematical relationship formulated in 1900 by German physicist Max Planck (1858–1947) to explain the spectral energy distribution of radiation emitted by a black body (a body that entirely absorbs all radiation energy that approaches it, reaches an equilibrium temperature, and then re-emits that energy as quickly as it absorbs it). Planck hypothesized that the sources of radiation are atoms in a state of oscillation and that the vibrational energy of each oscillator may have any of a series of discrete, or quantum, values but never any value in between. He further stated that when an oscillator changes from a state of energy E_1 [J] to a state of lower energy E_2 [J], the discrete amount of change in energy $E_1 - E_2$ [J], or quantum of radiation, is equal to the product of the frequency of the radiation v [Hz] and a constant $h = 6.62607015 \times 10^{-34}$ Js, now known as Planck's constant, that he determined from black body radiation data via

$$E_1 - E_2 = hv . \tag{13.15}$$

Planck's law for the energy density distribution $E_{\lambda,b}$ [Wm^{-2}m^{-1}] radiated per unit volume by a cavity of a black body in the wavelength interval λ to $\lambda + \Delta\lambda$ [m] (with $\Delta\lambda$ signifying an increment of wavelength) can be written in terms of Planck's constant h [Js], the speed of light c [ms^{-1}], the Boltzmann constant $k = 1.38064852 \times 10^{-23}$ m^{2}kgs^{-2}K^{-1}, and the absolute temperature T [K]:

$$E_{\lambda,b}(\lambda, T) = \frac{2\pi hc^2}{\lambda^5} \frac{1}{\exp\left(\frac{hc}{kT\lambda}\right) - 1} . \tag{13.16}$$

For a black body at temperatures up to many hundred degrees most of the radiation is in the infrared radiation region of the electro-magnetic spectrum. At

Radiating surface ϵ

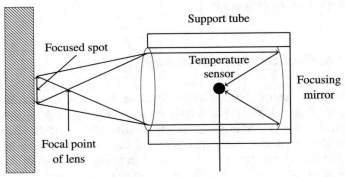

Fig. 13.13 Schematic of a basic radiometer

higher temperatures, the total radiated energy increases, and the intensity peak of the emitted spectrum shifts to shorter wavelengths so that a significant portion is radiated as visible light and beyond. The energy density distribution can be integrated over wavelength to provide the total emissive power of a black body using the Stefan–Boltzmann law

$$E_b(T) = \int_0^\infty E_{\lambda,b}(\lambda, T)d\lambda = \sigma T^4 \,, \qquad (13.17)$$

where $\sigma = 5.67 \times 10^{-8}$ Wm^{-2}K^{-4} is the Stefan–Boltzmann constant . For a non-black body, the emissive power is relative to the emissivity ϵ [–] of the body via

$$\epsilon = \frac{E}{E_b} \,. \qquad (13.18)$$

Therefore, it is important to know the emissivity of a surface when performing radiometry and pyrometry. A simple radiometer (pyrometer) is demonstrated in Fig. 13.13. The lens collects the radiation emitted from the radiating surface with emissivity ϵ [–]. The radiation is reflected and focused on a temperature sensor. The equilibrium temperature of the sensor is a direct measure of the magnitude of the radiation absorbed. The relationship between the temperature of the sensor and the radiation absorbed can be established by calibration experiments. Also, knowing the emissivity of the surface it is possible to estimate the temperature of the radiating surface.

Infrared thermal imaging is a popular radiometric technique to determine surface temperature of an object with emissivity ϵ. Many cameras are manufactured that count photons incident on a detector, and therefore they directly measure the incoming radiation to the sensor. Uncooled thermal cameras are those that do not regulate the temperature of the detector [10]. These cameras are physically lighter, inexpensive, and require less power to operate as compared to cooled thermal cameras, which regulate the detector temperature [11, 12]. Cooled thermal cameras are more accurate (± 1 K), while uncooled thermal cameras are less accurate (± 5 K) and should be often calibrated, taking into account the ambient temperature and the temperature of the detector [13].

The radiation energy emitted by the object, whose temperature is to be determined, that is being detected by photon counting U_{obj} [–] can be related to the object temperature T_{obj} [K] via a variation of the Planck's law given by:

$$U_{obj} = \frac{R}{\exp\left(\frac{B}{T_{obj}}\right) - F} - O , \qquad (13.19)$$

where R, B, F, O are constants of the camera to be determined by calibration under certain operating conditions (e.g. ambient or sensor temperatures). Once these constants are known, this formula can be inverted to give the object temperature via

$$T_{obj} = \frac{B}{\ln\left(\frac{R}{U_{obj}+O}\right) + F} . \qquad (13.20)$$

In thermal imaging, the radiation signal from the object U_{obj} [–] should be separated from the total radiation signal U_{tot} [–] that is recorded by the thermal camera, reflected radiation signal from the object U_{ref} [–], and the radiation signal from the atmosphere U_{atm} [–]. If τ [–] is the transmissivity of the atmosphere, the radiation signals discussed above are related by [13]:

$$U_{tot} = \epsilon\tau U_{obj} + \tau(1 - \epsilon)U_{ref} + (1 - \tau)U_{atm} . \qquad (13.21)$$

Many practical calculations assume $\tau = 1$ [–] [14], thus eliminating the last term above. Other assumptions are also made to treat the second term. Finally the radiation signal from the object U_{obj} [–] is determined using corrections to estimate the temperature of the object T_{obj} [K].

Microwave radiometry has been used successfully to measure vertical profiles of atmospheric temperature and humidity, which are important drivers of atmospheric turbulence, from near surface level up to an altitude of about 10 km [15, 16]. In addition, Doppler RAdio Detection And Ranging (RADAR) has been used to derive the vertical-wind field in the atmosphere at a scale of 1–2 km by performing vertical scans in the plane of the mean wind and tracking turbulent features on a scale of several hundred metres in the radial-velocity field between consecutive scans [17].

13.8   Light Detection And Ranging (LiDAR)

Light Detection And Ranging (LiDAR) is a surveying method that measures distance to a target by illuminating the target with laser light and measuring the reflected signal with a sensor. The signature of the return signal can then be used to make digital three-dimensional representations of the target. LiDARs have terrestrial, airborne, and mobile applications.

LiDAR systems are used to perform a range of measurements that include profiling clouds, measuring winds, studying aerosols, and quantifying various atmospheric components. Atmospheric components can in turn provide useful information including surface pressure (by measuring the absorption of oxygen or nitrogen), greenhouse gas emissions (carbon dioxide and methane), photosynthesis (carbon dioxide), fires (carbon monoxide), and humidity (water vapour). Atmospheric LiDAR remote sensing works in two ways: (1) by measuring backscatter from the atmosphere and (2) by measuring the scattered reflection off the ground (when the LiDAR is airborne) or other hard surfaces.

Backscatter from the atmosphere directly gives a measure of clouds and aerosols. Other derived measurements from backscatter such as winds or cirrus ice crystals require careful selection of the wavelength and/or polarization detected. Doppler LiDAR and Rayleigh Doppler LiDAR are used to measure temperature and/or wind speed along the beam by measuring the frequency of the backscattered light. The Doppler broadening of gases in motion allows the determination of properties via the resulting frequency shift. Scanning LiDARs have been used to measure atmospheric wind velocity.

Doppler LiDAR systems are also now beginning to be successfully applied in the renewable energy sector to acquire wind speed, turbulence, wind veer, and wind shear data. Both pulsed and continuous wave systems are being used. Pulsed systems use signal timing to obtain vertical distance resolution, whereas continuous wave systems rely on detector focusing.

For turbulence measurements in the atmosphere, LiDARs actually interact with atmospheric constituents such as particulates or fine water droplets that enable mapping three-dimensional wind fields at much higher frequency than SODARs, given the fact that light travels much faster than sound. Nevertheless, the sampling frequency of a LiDAR is limited by speed and efficiency of the software algorithms that convert the light signal to velocity measurements [18, 19].

Figure 13.14 shows the basic principle of a coherent continuous wave LiDAR system to measure the wind velocity. A beam of coherent radiation is transmitted towards a target, usually atmospheric aerosols, and a small fraction of the light is backscattered into a receiver. The motion of the aerosols along the beam direction results in a change in light's frequency δv [Hz] according to the Doppler shift. This frequency shift can be accurately measured by combining the return signal with a portion of the reference beam and sensing the resulting beats at the difference frequency on a photodetector. The atmospheric aerosols can be naturally occurring

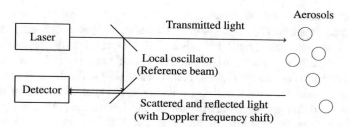

Fig. 13.14 Schematic of a bistatic LiDAR system

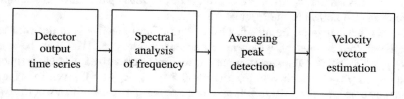

Fig. 13.15 Generic coherent continuous wave LiDAR signal processing

or anthropogenically generated. The composition may include elements of dust, organic matter, soot, or water, but the make-up does not really matter.

The local oscillator fulfils various roles. It defines the region of space from which light must be scattered. It allows for the rejection of other radiation sources in the environment, such as sunlight. It also provides a reference frequency of the light to allow a very accurate and precise velocity measurement. In addition, the local oscillator amplifies the return signal via the beating process to improve the sensitivity of the measurement. Unlike pulsed LiDARs, which use time of flight for the beam to travel into and back from an aerosol to determine velocity, coherent continuous wave LiDARs achieve operation at a given range by beam focusing. In other words, the beam of light is focused on a number of locations far away from the LiDAR to profile the wind environment.

Figure 13.15 shows the general signal processing scheme for the coherent continuous wave LiDAR. The detector output time series, which contains the beat signal information embedded in the broadband noise, is digitized using an Analogue-to-Digital Converter (ADC). Then spectral analysis of the time series is performed using Fast Fourier Transform (FFT) to generate the Doppler spectra. Typically such spectra is averaged using multiple measurements to improve the Signal-to-Noise Ratio (SNR). This allows for the identification of the Doppler peak. The frequency of this peak, in comparison to the frequency of the Laser can then be analysed using an algorithm to determine the velocity component in the direction of the beam. Depending on the speed of the measurement and calculation of velocity, both mean and turbulence statistics of the wind velocity vector can be computed. To measure all components of the wind velocity vector, none of the three LiDAR beams should be perpendicular to the wind flow, or else no Doppler shifting will occur.

The identification of the Doppler shift can be elaborated further using basic mathematical relations [20, 21]. Assume that the return spectrum is only marginally broadened by atmospheric motion. Since LiDARs essentially measure digitized signals, the peak of the spectrum is located only within one discrete spectral channel. This is referred to as the Discrete Spectral Peak (DSP). The LiDAR return signal has a power spectrum characterized by its ensemble-averaged spectral density. It can be assumed that the spectrum does not change during the period of observation, which necessitates the sampling period to be less than the timescale of atmospheric fluctuations. Suppose that the sampling frequency is F_s [Hz], in which case the Doppler shifts can be identified from a frequency of zero up to half of this sampling frequency, also known as the Nyquist frequency.

We can normalize all frequencies by F_s [Hz]. Our objective is to determine the Doppler-shifted frequency, which after normalization by F_s [Hz], can be shown as f_1 [–]. The FFT of the signal results in multiple normalized discrete frequency components f_i [–], which would satisfy $(-0.5 \leq f_i < +0.5)$, given the Nyquist frequency consideration. There will be a noise power spectrum, which would always be present in the background, even without a return signal. This spectrum can be written as $\phi_N(f_i)$, with appropriate units. The LiDAR return power spectrum is conditional on f_1 [–] and can be normalized by the noise spectrum at the same frequency, $\phi_N(f_i)$; therefore, it can be shown as $\phi_S(f_i|f_1)$ [–], which is now a unitless quantity. With this convention, $\phi_S(f_i|f_1)$ [–] is also the SNR [–] within a single spectral channel. For a single return data sample, $\phi_S(f_i|f_1)$ or SNR [–] can be written as δ [–].

It has been demonstrated that the return normalized power spectrum is best approximated by a Gaussian profile, but it is superimposed on the normalized noise power spectrum, B, whose magnitude is 1 over all measured channels. Therefore, the measured normalized total power spectrum and that of the signal will be [21]

$$\phi(f_i|f_1) = \phi_S(f_i|f_1) + B , \tag{13.22}$$

$$\phi_S(f_i|f_1) = \frac{\delta}{\sqrt{2\pi}\, f_2} \exp\left(-\frac{(f_i - f_1)^2}{2 f_2^2}\right) , \tag{13.23}$$

where $B = 1$ for $-0.5 \leq f_i < +0.5$ and f_2 is the second moment width, i.e. the square root of the second central moment of the signal component $\phi_S(f_i|f_1)$.

For a single, discretely sampled, measurement from a single LiDAR return, a periodogram is obtained, which is comprised of the signal plus noise for M [–] spectral channels

$$\mathbf{X} = (X_1, X_2, \ldots, X_M) . \tag{13.24}$$

If n [–] samples are accumulated (same idea as averaged), then the spectra will be added, or accumulated, for each channel. In this case the sample spectrum will be given by:

$$\mathbf{x} = (x_1, x_2, \ldots, x_M) = \sum_{j=1}^{n} \mathbf{X}_j ,$$ (13.25)

where each component of this sample is obtained by $x_i = \sum_{j=1}^{n}(X_i)_j$. A successful LiDAR system would infer the Doppler-shifted frequency f_1 from such a spectrum using advanced statistical techniques for parameter estimation, such as the Maximum Likelihood (ML) technique to determine the DSP. A successful LiDAR system would attempt to maximize the SNR or δ [–] with appropriate choices of n and M [–], for a given spectral distribution broadening characterized by f_2. For accumulating separate discretely sampled spectra, it has been shown that the product nM [–] is related to δ [–] such that [21]

$$nM = \frac{64}{\delta^{\frac{4}{3}}} .$$ (13.26)

This equation can be used to determine approximately what order of accumulation n [–] is required for M [–] spectral channels to achieve a desired SNR or δ [–].

The estimation techniques for separate discretely-sampled spectra are very computationally expensive and can limit the spatial and temporal resolutions of measuring wind velocity vector components in real time. Therefore, other advanced techniques can be deployed to enhance measurement resolutions. An alternative to formation of the periodogram from the time series data sample is the formation of the sample autocorrelation function, or correlogram. The correlogram consists of a set of autocorrelation estimates for some number of time lags obtained directly from the data sample. Correlograms can be computed and manipulated much more computationally efficiently, and combined with other parameter estimation techniques, other approaches enable real time measurement of wind velocity vector components at greater spatial and temporal resolutions [20].

Figure 13.16 shows the coordinate system of a generic coherent continuous wave LiDAR. The LiDAR scans air parcels at the surface of an inverted cone with a fixed zenith angle φ [Rad] and a variable scanning angle $\phi_n \in [0, 2\pi]$ [Rad] for N [–] choices (at least 3) with respect to the x axis (East). At each height z [m], a complete scan revolution may take up to 3 s. At the instant shown, beam n [–] can be characterized with a unit vector given by:

$$\mathbf{B}_n = \sin \varphi \cos \phi_n \mathbf{i} + \sin \varphi \sin \phi_n \mathbf{j} + \cos \varphi \mathbf{k} .$$ (13.27)

This beam measures the projection of the wind velocity vector $\mathbf{U} = (U, V, W)$ [ms^{-1}] along $\mathbf{B}_n$ [–], i.e. U_{B_n} [ms^{-1}]. With only one beam, the three components of the wind velocity vector cannot be distinguished. However, with a minimum of three beams, the components can be inferred. In fact the system of equations below

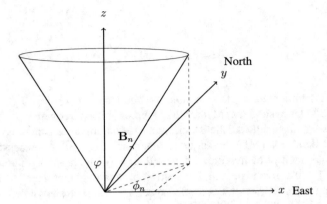

Fig. 13.16 Coordinate system of a generic coherent continuous wave LiDAR

can be inverted to provide the wind velocity components after fitting of data

$$\begin{bmatrix} U_{B_1} \\ \vdots \\ U_{B_N} \end{bmatrix} = \begin{bmatrix} \mathbf{U}.\mathbf{B}_1 \\ \vdots \\ \mathbf{U}.\mathbf{B}_N \end{bmatrix} = \begin{bmatrix} \sin\varphi\cos\phi_1 & \sin\varphi\sin\phi_1 & \cos\varphi \\ \vdots & \vdots & \vdots \\ \sin\varphi\cos\phi_N & \sin\varphi\sin\phi_N & \cos\varphi \end{bmatrix} \begin{bmatrix} U \\ V \\ W \end{bmatrix} . \quad (13.28)$$

Problems

13.1 For long wavelengths, the refractive index of air can be given as a function of pressure P [Pa] and temperature T [K], as well as a reference pressure $P_0 = 101.3\,\text{kPa}$ and temperature $T_0 = 300\,\text{K}$, such that

$$n(P, T) = 1 + 0.000293 \frac{P}{P_0} \frac{T_0}{T} . \quad (13.29)$$

An engineer wants to use this relationship to design a Schlieren system for imaging the thermal plumes associated with the human body. The engineer wishes to have an estimate of light beam deflection ϵ_y [m m^{-1}]. Suppose that the optical extent of the Schlieren system is $L = 10\,\text{m}$. The human body can be assumed to be situated in an environment with the reference pressure and temperature noted above, i.e. P_0 [Pa] and T_0 [K], respectively. Further, we can assume that the human body creates thermal gradients as high as $\Delta T = 10\,\text{K}$ above the reference temperature over a distance $\Delta y = 0.1\,\text{m}$, but it does not significantly alter the reference pressure in the environment, i.e. the pressure P [Pa] is close to P_0 [Pa] everywhere surrounding the body. Help the engineer estimate ϵ_y [m m^{-1}] for this setup. Hint: assume that $n_0 = n(P_0, T_0)$ [–] and that the average gradient of the refractive index can be given by:

$$\frac{\overline{\partial n}}{\partial y} \approx \frac{n(P_0, T_0) - n(P_0, T_0 + \Delta T)}{\Delta y} . \quad (13.30)$$

13.2 In an LDV experiment, a velocity of $U = 1\,\text{ms}^{-1}$ is measured. In this system the wavelength of the laser system is $\lambda = 632.8\,\text{nm}$, and the angle between the two intersecting light beams from the equal-path beam splitter is $\theta = \frac{\pi}{8}\,\text{Rad}$. Calculate the Doppler shift frequency f_D [Hz].

13.3 A small temperature probe with surface area A_p [m^2] is placed in a high temperature furnace, where the average fluid temperature is T_f=800°C, and the average internal surface temperature of the furnace is T_s=500°C. The emissivity of the probe is $\epsilon_p = 0.8$ [–], and the convective heat transfer coefficient between the fluid and probe is $h = 100\,\text{Wm}^{-2}\text{K}^{-1}$. Find the temperature of the probe T_p [K] assuming thermal equilibrium for the probe such that

$$hA_p\left(T_f - T_p\right) = A_p\epsilon_p\sigma\left(T_p^4 - T_s^4\right).\tag{13.31}$$

13.4 An engineer is designing a coherent continuous wave LiDAR system. She configures the system to have three beam directions, $\mathbf{B}_1$, $\mathbf{B}_2$, and $\mathbf{B}_3$ [–], i.e. $N = 3$ [–]. The system is required to measure the profile of the wind velocity vector at 10 altitudes up to 100 m. The system should estimate the profile of wind velocity vector components every hour, i.e. at a time resolution of 1 hr. Therefore, overall, she has to measure the wind velocity components at 30 points around the surface of the scan cone. She has two options to scan using this LiDAR: (1) she can shine the beam at a given point for 2 min and complete a measurement at that point, then she can shine the beam at the next point for 2 min to complete the measurement at the next point, and so on; (2) alternatively, she can measure a point once for a few seconds and move on to the next point, while after finishing one complete round covering all the 30 points, she can come back to the first point and measure again. Which option do you prefer? Why?

13.5 A LiDAR system is considered that processes separate discretely sampled spectra to identify the normalized Doppler-shifted frequency f_1 [–]. A scientist has two options to operate the system. In option 1, $n = 32$ accumulations of the signal are possible for $M = 64$ [–] spectral channels, while in option 2, $n = 16$ accumulations of the signal are possible for $M = 64$ spectral channels. Help the scientist calculate the Signal to Noise Ratio (SNR) or δ [–] for either option. Which option results in a higher SNR or δ [–]?

13.6 A Helium-neon laser (He-Ne lasers $\lambda = 633$ nm) is used for a PIV experiment with average particle diameter of $d_p = 1\,\mu\text{m}$. Using the Mie scattering theory, calculate the normalized diameter q [–], which provides the number of peaks for the scattered light intensity around the particle over the angular distribution around the particle from 0 to 180°.

References

1. Settles G S (2001) Schlieren and shadowgraphy techniques: visualizing phenomena in transparent media. Springer, Berlin
2. Aliabadi A A, Lim K W, Rogak S N et al (2011) Steady and transient droplet dispersion in an air-assist internally mixing cone atomizer. Atomization Spray 21:1009–1031
3. Dennis K, Siddiqui K (2021) Visualization and characterization of thermals in a heated turbulent boundary layer. Exp Therm Fluid Sci 120:110237
4. Dennis K, Siddiqui K (2021) The influence of wall heating on turbulent boundary layer characteristics during mixed convection. Int J of Heat Fluid Fl 91:108839
5. Dennis K (2020) Experimental investigation on the impact of wall heating on mixed convection turbulent boundary layer flow structure. Western University, London
6. Raffel M, Willert C E, Scarano F et al (2018) Particle image velocimetry: a practical guide. Springer, Berlin
7. Hong J, Toloui M, Chamorro L P et al (2014) Natural snowfall reveals large-scale flow structures in the wake of a 2.5-MW wind turbine. Nat Commun 5:4216
8. Hargather M J, Lawson M J, Settles G S et al (2011) Seedless velocimetry measurements by Schlieren image velocimetry. AIAA J 49:611–620
9. Drain L E (1980) The laser Doppler technique. John Wiley & Sons, New York
10. Colomina I, Molina P (2014) Unmanned aerial systems for photogrammetry and remote sensing: a review. ISPRS J Photogramm 92:79–97
11. Ribeiro-Gomes K, Hernández-López D, Ortega J F et al (2017) Uncooled thermal camera calibration and optimization of the photogrammetry process for UAV applications in agriculture. Sensors-Basel 17:2173
12. Rahaghi A I, Lemmin U, Sage D et al (2019) Achieving high-resolution thermal imagery in low-contrast lake surface waters by aerial remote sensing and image registration. Remote Sens Environ 221:773–783
13. Byerlay R A E, Nambiar M K, Nazem A et al (2020) Measurement of land surface temperature from oblique angle airborne thermal camera observations. Int J Remote Sens 41:3119–3146
14. Usamentiaga R, Venegas P, Guerediaga J et al (2014) Infrared thermography for temperature measurement and non-destructive testing. Sensors-Basel 14:12305–12348
15. Candlish L M, Raddatz R L, Asplin M G et al (2012) Atmospheric temperature and absolute humidity profiles over the Beaufort Sea and Amundsen Gulf from a microwave radiometer. J Atmos Ocean Tech 29:1182-1201
16. Ramamurthy P, González J, Ortiz L et al (2017) Impact of heatwave on a megacity: an observational analysis of New York City during July 2016. Environ Res Lett 12:054011
17. Hogan R J, Illingworth A J, Halladay K (2008) Estimating mass and momentum fluxes in a line of cumulonimbus using a single high-resolution Doppler radar. Q J Roy Meteor Soc 134:1127–1141
18. Huang M, Gao Z, Miao S et al (2017) Estimate of boundary-layer depth over Beijing, China, using Doppler LiDAR data during SURF-2015. Bound-Lay Meteorol 162:503–522
19. Halios C H, Barlow J F (2018) Observations of the morning development of the urban boundary layer over London, UK, taken during the ACTUAL Project. Bound-Lay Meteorol 166:395–422
20. Rye B J, Hardesty R M (1993) Discrete spectral peak estimation in incoherent backscatter heterodyne LiDAR. II. Correlogram accumulation. IEEE T Geosci Remote 31:28–35
21. Rye B J, Hardesty R M (1993) Discrete spectral peak estimation in incoherent backscatter heterodyne lidar. I. Spectral accumulation and the Cramer-Rao lower bound. IEEE T Geosci Remote 31:16–27

Part III
Turbulence Modelling and Simulation

Chapter 14
Introduction to Modelling and Simulation

Abstract This chapter lays the foundation for fluid modelling and simulations in general, with a particular focus on turbulence. Three main paradigms in turbulence modelling and simulation are introduced. These are Reynolds-Averaged Navier–Stokes (RANS) models, Large-Eddy Simulations (LES), and Direct Numerical Simulations (DNS). The concept of model or simulation completeness is defined. The turbulence model or simulation closure problem is revisited. The advent of digital computation in relation to fluid modelling and simulations is discussed. The idea of the *flops* (floating-point operations per second) is introduced as a measure of the speed of computation. Different digital computing paradigms are discussed, involving High Performance Computing (HPC), Central Processing Units (CPUs), and Graphics Processing Units (GPUs).

14.1 Preliminaries

More than one century of experience has shown that the *turbulence problem* is inconveniently difficult. Though conceptually simple, the turbulence problem is the unsolved problem of classical physics. In other words, there is no simple analytical theory that completely describes physics of turbulence. Instead, engineers and scientists rely on the ever-increasing power of digital computers to *model* or *simulate* turbulence for a given application to calculate the relevant properties of turbulent flows. Modelling or simulating turbulence *exactly* is still out of reach for practical problems given today's computational power accessible to engineers and scientists. Instead, a variety of models and simulation tools are accessible that calculate few or many relevant properties of turbulent flows.

It is worth distinguishing the technical difference between a turbulence *model* and a turbulence *simulation*. In a turbulence model, equations are solved to give some *mean* quantities, for instance mean velocity or Reynolds stress. In contrast, in a turbulence simulation, equations are solved for time-dependent properties of flow for a particular realization of the flow. Given enough realizations, it is still possible to derive mean quantities by performing statistical analysis on simulation results in the post-processing phase as opposed to the solution calculation phase.

Nevertheless, the terms *model* and *simulation* are often freely interchanged in the literature, so it is the responsibility of the careful reader to infer whether an approach actually refers to a model or a simulation.

14.2   Summary of Approaches

Turbulent viscosity models are among the most computationally affordable, but not necessarily conceptually or mathematically simple, models. As a class of Reynolds-Averaged Navier–Stokes (RANS) models, turbulent viscosity models calculate mean properties of the flow, such as velocity or Reynolds stress. Two common examples of such models are the *mixing-length* and the $k - \epsilon$ models.

In Large-Eddy Simulation (LES), equations are solved for *filtered* properties of the flow, such as velocity, which are actually resolved and give a realization of the flow at the larger scale of turbulent motion. Smaller scales of motion, however, are not resolved but modelled. LES requires significantly higher computational resources but provides more accurate results and is practical for many problems. The rationale behind suitability of LES is that for many applications the physics of the flow is dominated by large-scale fluctuations, while smaller-scale fluctuations play a lesser significant role and do not have to be resolved or realized. With exponential increase in computational power, LES has gained popularity in the recent decades and has been applied to many practical engineering and science problems.

In Direct Numerical Simulation (DNS), equations are solved for properties of the flow, such as velocity, which are resolved and give a realization of the flow across all spatial and temporal scales. DNS is computationally very expensive and not practical for most problems. However, it can calculate all turbulence properties of the flow. The computational requirements for DNS are so prohibitive that it is still primarily used as a research tool for fundamental studies [1].

14.3   Model or Simulation Completeness

A turbulence model or simulation is termed *complete* if its constituent equations are free from flow-dependent specifications. Such specifications include material properties (density and viscosity), initial and boundary conditions, and numerical discretization. For example, DNS and the $k - \epsilon$ turbulent viscosity model are complete since their equations do not depend on flow-dependent specifications. On the other hand LES and the mixing-length models are incomplete. LES results depend on numerical discretization and the mixing-length model needs to be specified by a mixing-length, both of which are flow dependent.

14.4   Turbulence Model or Simulation Closure Problem

As noted in Chap. 4, when modelling or simulating turbulence, most often the number of unknowns in the set of equations for turbulent flow is larger than the number of equations. When one tries to include new equations to balance the number of unknowns and equations, one finds out that the number of unknowns further increases so that, again, the number of unknowns is larger than the number of equations considered. This is called the *closure problem* and is attributed to the non-linear characteristics of the governing equations. One approach is to use only a finite number of equations, i.e. stop writing down more equations, and then approximate the remaining unknowns in terms of the known quantities [2].

The *art* of modelling and simulating turbulence is to choose or develop an approach that utilizes the least number of equations and unknowns but provides a reasonably acceptable solution for a given problem. It is also known that more equations and unknowns do not necessarily guarantee a more acceptable solution for a given problem. For instance, it has been shown that for planetary boundary-layer atmospheric flows, only a few equations suffice to arrive at an acceptable solution [3]. The proper choice or development of a suitable model or simulation tool usually follows from years of practice and experience in the field.

14.5   Digital Computation

There has been significant advancement in speed and volume of calculations enabled by digital computers over the last fifty years. The amount of computation in *flops* (floating-point operations per second) over a specified time is a measure of a computer's ability to perform calculations. For instance a processor technology is specified by the number of gigaflops, teraflops, or petaflops per second.

High Performance Computing (HPC) hardware configurations have been developed on Central Processing Units (CPU) and Graphics Processing Units (GPU) for the purpose of turbulence modelling and simulation. Both homogeneous and heterogeneous hardware architectures have been considered: in the homogeneous configuration only multiple CPUs are considered, while in the heterogeneous configuration both CPUs and GPUs are integrated. It is known that a GPU consists of thousands of computational cores that can, theoretically, perform arithmetic computations faster than a multi-threaded (parallel) CPU. However, a GPU requires time for communicating data with the main computer memory. This factor slows down the speed of overall computation. On the other hand, a multi-threaded CPU has direct access to the computer memory and benefits from fast data transfer but performs arithmetic computations more slowly than a GPU. As a trade-off, most recent research efforts to accelerate models and simulations have considered using very few GPUs to off-load the *solver* module of the model or simulation from

CPU, i.e. the part of the code that solves a linear system of equations and is most computationally expensive [4]. This is also known as GPU-accelerated computing.

Problems

14.1 Describe the difference between turbulence modelling and turbulence simulation.

14.2 What are flops?

14.3 Describe the pros and cons of CPUs and GPUs.

14.4 Most recently and alternative to CPUs and GPUs, Field-Programmable Gate Arrays (FPGAs) have been used for digital computation. FPGAs contain an array of programmable logic blocks and a hierarchy of reconfigurable interconnects that allow the blocks to be wired together. Logic blocks can be configured to perform simple or complex logic operations (e.g. AND, OR, etc.). In most FPGAs, logic blocks also include memory elements, which may be simple flip-flops or more complete blocks of memory. Although programmable to some degree, an FPGA does not function similar to a processor, so it cannot run any general program stored in the memory. Rather, an FPGA is meant to perform specific computational tasks or algorithms for a particular application. However, because of its parallel and reconfigurable architecture, an FPGA can run logic operations much faster and more energy efficiently than CPUs or GPUs. Discuss ways in which an FPGA can be used to accelerate a turbulence model or simulation.

References

1. Pope S B (2000) Turbulent flows. Cambridge University Press, Cambridge
2. Stull R B (1988) An introduction to boundary layer meteorology. Kluwer Academic Publishers, Dordrecht
3. Mellor G L, Yamada T (1974) A hierarchy of turbulence closure models for planetary boundary layers. J Atmos Sci 31:1791–1806
4. Codyer, S R, Raessi M, Khanna G (2012) Using graphics processing units to accelerate numerical simulations of interfacial incompressible flows. Fluids Engineering Division Summer Meeting - Volume 1: Symposia, Parts A and B:625–634

Chapter 15
Turbulent-Viscosity Models

Abstract This chapter introduces the turbulent-viscosity models. In this modelling paradigm the turbulence mechanisms of fluid transport are modelled as a viscous process with the assumption of a turbulent viscosity. The gradient-diffusion hypothesis is at the centre of such modelling paradigm. A series of turbulent-viscosity models are introduced, involving the algebraic models, the Spalart–Allmaras model, the turbulence kinetic energy models, the $k - \epsilon$ model, and the $k - \omega$ model. Finally, turbulent-viscosity models are introduced for the atmospheric boundary layer.

15.1 Preliminaries

The methodology developed in this chapter is adapted from [4]. *Turbulent-viscosity models* are based on the *turbulent-viscosity hypothesis*, introduced using Eq. 4.22, in which the Reynolds stresses are given as

$$\langle u_i u_j \rangle = \frac{2}{3} k \delta_{ij} - v_T \left(\frac{\partial \langle U_i \rangle}{\partial x_j} + \frac{\partial \langle U_j \rangle}{\partial x_i} \right) . \tag{15.1}$$

These models are a class of *Reynolds-Averaged Navier–Stokes (RANS)* models, in which the transport equations are solved for the mean velocity field. In a simple shear flow, the shear stress is given by:

$$\langle uv \rangle = -v_T \frac{\partial \langle U \rangle}{\partial y} , \tag{15.2}$$

in which u [ms^{-1}] is velocity fluctuation parallel to a wall in the x direction, and v [ms^{-1}] is velocity fluctuation normal to the wall in the y direction. As was described by Eq. 4.24, v_T [m^{2}s^{-1}] is not a constant but a field that depends on both position and time, i.e. $v_T(\mathbf{x}, t)$ [m^{2}s^{-1}]. If this field can be modelled conveniently, this provides a convenient closure scheme for turbulence modelling. Turbulent viscosity can be written as the product of a velocity scale $u^*(\mathbf{x}, t)$ [ms^{-1}] and lengthscale

$\ell^*(\mathbf{x}, t)$ [m], both of which are fields themselves

$$v_T = u^*\ell^* . \tag{15.3}$$

This is analogous to the kinetic theory of gasses, with the kinematic viscosity given by:

$$\nu \sim \frac{1}{2}\overline{C}\lambda , \tag{15.4}$$

where $\overline{C}$ [ms^{-1}] is the mean molecular speed and λ [m] is the mean free path [1]. The main task in turbulent-viscosity models is to specify these velocity and lengthscales. In mixing length models, ℓ^* [m] is specified on the basis of the geometry of the flow. In one-equation turbulence kinetic energy models, ℓ^* [m] is still specified on the basis of the geometry of the flow, while u^* [ms^{-1}] is specified by the turbulence kinetic energy k [m^{2}s^{-2}], for which a transport equation is solved. In two-equation models, e.g. the $k - \epsilon$ model, both ℓ^* [m] and u^* [ms^{-1}] are expressed as functions of k [m^{2}s^{-2}] and dissipation rate ϵ [m^{2}s^{-3}], for which transport equations are solved.

The *turbulent-viscosity hypothesis* is based on two assumptions. First, the intrinsic assumption is that at each point and time the Reynolds stress *anisotropy* $a_{ij} \equiv \langle u_i u_j \rangle - \frac{2}{3}k\delta_{ij}$ [m^{2}s^{-2}] (Eq. 4.12) is determined by the mean velocity gradients $\partial\langle U_i \rangle/\partial x_j$ [s^{-1}]. Second, there is the specific assumption that the relationship between a_{ij} [m^{2}s^{-2}] and $\partial\langle U_i \rangle/\partial x_j$ [s^{-1}] is

$$\langle u_i u_j \rangle - \frac{2}{3}k\delta_{ij} = -v_T \left(\frac{\partial\langle U_i \rangle}{\partial x_j} + \frac{\partial\langle U_j \rangle}{\partial x_i} \right) , \tag{15.5}$$

given before in Eq. 4.22. In short form, we can write

$$a_{ij} = -2v_T \overline{S}_{ij} , \tag{15.6}$$

where $\overline{S}_{ij}$ [s^{-1}] is the *mean rate-of-strain tensor*, already defined using Eq. 2.17. This concept is directly analogous to the relation for the viscous stress in a Newtonian fluid

$$- (\tau_{ij} + P\delta_{ij})/\rho = -2\nu S_{ij} . \tag{15.7}$$

Related to the concept of turbulent-viscosity hypothesis is the *gradient-diffusion hypothesis* introduced earlier using Eq. 4.19 as

$$\langle \mathbf{u}\phi' \rangle = -\Gamma_T \nabla\langle\phi\rangle , \tag{15.8}$$

which states that the scalar flux $\langle \mathbf{u}\phi' \rangle$ [ms^{-1}] is aligned with the mean scalar gradient. It was stated earlier that Γ_T [m^{2}s^{-1}] is *turbulent diffusivity* and should not be confused with *molecular diffusivity*.

The following sections present the turbulent-viscosity models developed and ordered with increasing level of sophistication and accuracy. From historical point of view, it must be understood that the turbulence modelling field made significant progress in the recent decades albeit slowly given the difficulty of the problem. Therefore, the reader is encouraged to appreciate all the developments that continue to have applications today given the level of sophistication and accuracy required for a problem.

15.2 Algebraic Models

The *algebraic models* are classified into *uniform turbulent-viscosity models* and *mixing length models*. The mixing length model is frequently used, particularly in simple shear flows over flat surface in atmospheric and oceanic applications. In this model, the *mixing length* $\ell_m(x, y)$ [m] is specified as a function of position, and then the turbulent viscosity is obtained as

$$v_T = \ell_m^2 \left| \frac{\partial \langle U \rangle}{\partial y} \right| , \tag{15.9}$$

where $\langle U \rangle$ [ms^{-1}] is mean velocity parallel to the surface and y is the direction normal to the surface. In the *log-law region* introduced in Chap. 5, mixing length and turbulent viscosity can be specified by:

$$\ell_m = \kappa y , \tag{15.10}$$

$$v_T = u_\tau \kappa y , \tag{15.11}$$

where κ [$-$] is the von Kármán constant [2]. Figure 15.1 shows the mixing length relationship with distance from the wall, in addition to the velocity profile near the wall in the *log-law region*. In this region there is a linear relationship between the mixing length and distance from the wall.

Mixing length models can be applied for generalized flows with the following formulation:

$$v_T = \ell_m^2 (2\overline{S}_{ij}\overline{S}_{ij})^{1/2} = \ell_m^2 S , \tag{15.12}$$

where $\overline{S}_{ij}$ [s^{-1}] is the *mean rate-of-strain tensor*, already defined using Eq. 2.17. Other generalized flows formulate the mixing length model with

$$v_T = \ell_m^2 (2\overline{\Omega}_{ij}\overline{\Omega}_{ij})^{1/2} = \ell_m^2 \Omega , \tag{15.13}$$

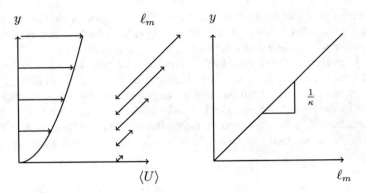

Fig. 15.1 Specification of the mixing length in the *log-law region* near the wall

where $\overline{\Omega}_{ij}$ [s^{-1}] is the *mean rate-of-rotation tensor*, already defined using Eq. 2.18. Note that in these models, ℓ_m [m] must still be specified given some information about the geometry of the flow. As a general rule, mixing length approaches zero at walls or surfaces and reaches some maximum in the interior of the flow or far away from walls.

15.3   Spalart–Allmaras Model

The next level of sophistication in turbulence modelling is possible by considering a transport equation for the turbulent viscosity itself. This is motivated by the fact that, after all, turbulent viscosity is not constant but a field that varies spatiotemporally. Therefore, its transport must be governed by an equation. Reference [3] proposed a one-equation model, although with similar preceding proposals in the literature, developed for aerodynamic applications, where a single model transport equation is solved for turbulent viscosity v_T [m^{2}s^{-1}]. The equation is given as

$$\underbrace{\frac{\overline{D}v_T}{\overline{D}t}}_{\text{Mean Substantial Derivative}} \equiv \underbrace{\frac{\partial v_T}{\partial t}}_{\text{Storage}} + \underbrace{\langle \mathbf{U} \rangle . \nabla v_T}_{\text{Advection}}$$

$$= \underbrace{\nabla . \left(\frac{v_T}{\sigma_v} \nabla v_T \right)}_{\text{Turbulent-Viscosity Flux Divergence}} + \underbrace{S_v}_{\text{Source/Sink}} \quad , \tag{15.14}$$

where σ_v [−] is the Prandtl number for turbulent viscosity and S_v [m^{2}s^{-2}] is the source or sink term that depends on many variables such as laminar viscosity v [m^{2}s^{-1}], turbulent viscosity v_T [m^{2}s^{-1}], mean vorticity or rate of rotation Ω [s^{-1}], turbulent-viscosity gradient $\nabla \nu_T$ [ms^{-1}], and distance to the nearest wall ℓ_w [m]. This model is successful for aerodynamic flows but has limitations as a general model. For instance it cannot account for decay of turbulent viscosity in isotropic turbulence [4].

15.4 Turbulence Kinetic Energy Models

It was introduced earlier that turbulent viscosity can be written as $\nu_T = u^* \ell^*$ [m^{2}s^{-1}], i.e. the product of a velocity scale and a lengthscale. In the mixing length model $\ell^* = \ell_m$ [m], and the velocity scale was given as

$$u^* = \ell_m |\frac{\partial \langle U \rangle}{\partial y}| \, . \tag{15.15}$$

This approximation requires that u^* [ms^{-1}] be zero wherever $|\partial \langle U \rangle / \partial y|$ [s^{-1}] is zero. This approximation is far from reality for many cases where $|\partial \langle U \rangle / \partial y|$ [s^{-1}] is zero but the velocity scale u^* [ms^{-1}] is not necessarily zero. It has been suggested by [5] and [6], independently, that the velocity scale can be better formulated by the turbulence kinetic energy, i.e.

$$u^* = c k^{1/2} \, , \tag{15.16}$$

where c [−] is a constant. If the lengthscale is again taken to be the mixing length, then the turbulent viscosity can be formulated as

$$\nu_T = c \ell_m k^{1/2} \, . \tag{15.17}$$

The constant c [−] can be fitted for various turbulent regimes. For instance $c \approx 0.55$ [−] yields a correct approximation in the log-law region. This formulation requires knowledge of $k(\mathbf{x}, t)$ [m^{2}s^{-2}], i.e. the turbulence kinetic energy field. If we assume we have a knowledge of ℓ_m [m] or we can formulate it based on known quantities, we can then develop a transport equation for k [m^{2}s^{-2}], which can be solved for k [m^{2}s^{-2}]. References [5] and [6] suggested the following transport equation:

$$\underbrace{\frac{\overline{Dk}}{Dt}}_{\text{Mean Substantial Derivative}} \equiv \underbrace{\frac{\partial k}{\partial t}}_{\text{Storage}} + \underbrace{\langle \mathbf{U} \rangle . \nabla k}_{\text{Advection}}$$

$$= \underbrace{-\nabla . \mathbf{T}'}_{\text{Energy Flux Divergence}} + \underbrace{\mathcal{P}}_{\text{Production}} - \underbrace{\epsilon}_{\text{Dissipation}} \, . \tag{15.18}$$

Note that the production term $\mathcal{P}$ [m^{2}s^{-3}] is referred to as a mechanism that generates turbulence kinetic energy, and hence the term has a positive sign in the equation. On the other hand, the dissipation term ϵ [m^{2}s^{-3}] is responsible for consuming the kinetic energy, and hence the term has a negative sign. The total derivative $\overline{D}k/\overline{D}t$ [m^{2}s^{-3}] and production terms $\mathcal{P}$ [m^{2}s^{-3}] are in closed form, but the energy flux divergence $-\nabla.\mathbf{T}'$ [m^{2}s^{-3}] and dissipation terms ϵ [m^{2}s^{-3}] are unknowns and should be modelled or closed further.

It was discussed extensively in Chap. 8 that the dissipation rate ϵ [m^{2}s^{-3}] scales as u_0^3/ℓ_0 [m^{2}s^{-3}], where u_0 [ms^{-1}] and ℓ_0 [m] are the velocity scale and the lengthscale of the energy-containing motions. The same scaling can be used to formulate the dissipation rate

$$\epsilon = C_D \frac{k^{3/2}}{\ell_m} , \tag{15.19}$$

where C_D [$-$] is a model constant. Indeed, an examination of the log-law region yields that $C_D = c^3$ [$-$]. This modelling approach in fact eliminates ℓ_m [m] since it vanishes in the turbulent-viscosity equation

$$\nu_T = cC_D \frac{k^2}{\epsilon} . \tag{15.20}$$

The energy flux $\mathbf{T}'$ [m^{3}s^{-3}] can be modelled using the gradient-diffusion hypothesis such that

$$\mathbf{T}' = -\frac{\nu_T}{\sigma_k} \nabla k , \tag{15.21}$$

where σ_k [$-$] is the turbulent Prandtl number for turbulence kinetic energy, generally assumed to be one [4]. The energy flux divergence term in the turbulence kinetic energy equation accounts for the flux of k [m^{2}s^{-2}] down the gradient of k [m^{2}s^{-2}] due to velocity and pressure fluctuations. This term ensures that the resulting transport equation model for k [m^{2}s^{-2}] yields smooth solutions and that a boundary condition can be imposed on k [m^{2}s^{-2}] everywhere in the boundary of the domain. Otherwise the model may diverge if other transport mechanisms for k [m^{2}s^{-2}] are much smaller than this term. In summary, the one-equation model based on k [m^{2}s^{-2}] consists of the following transport equation and the closure schemes:

$$\underbrace{\frac{\overline{D}k}{\overline{D}t}}_{\text{Mean Substantial Derivative}} \equiv \underbrace{\frac{\partial k}{\partial t}}_{\text{Storage}} + \underbrace{\langle \mathbf{U} \rangle . \nabla k}_{\text{Advection}}$$

$$= \underbrace{\nabla . \left(\frac{\nu_T}{\sigma_k} \nabla k \right)}_{\text{Energy Flux Divergence}} + \underbrace{\mathcal{P}}_{\text{Production}} - \underbrace{\epsilon}_{\text{Dissipation}} , \tag{15.22}$$

$$v_T = ck^{1/2}\ell_m \, , \tag{15.23}$$

$$\epsilon = C_D \frac{k^{3/2}}{\ell_m} \, , \tag{15.24}$$

$$\ell_m(\mathbf{x}, t) \text{ known} \, . \tag{15.25}$$

This model belongs to a class of models known as *one-equation models*. These models are called so because only one extra equation has to be solved to close the turbulence model.

15.5   The $k - \epsilon$ Model

The $k - \epsilon$ model is among the most popular turbulence models, particularly within the engineering community and comprehensively cited in the 1970s [7]. This model belongs to a class of models known as *two-equation models* where two extra equations have to be solved to close the turbulence model. The two unknowns in this case are turbulence kinetic energy k $[\text{m}^2\text{s}^{-2}]$ and dissipation rate ϵ $[\text{m}^2\text{s}^{-3}]$. From these two unknowns and equations other quantities can be formed. For instance a lengthscale in turbulence can be formulated as $L = k^{3/2}/\epsilon$ [m], a timescale can be formulated as $\tau = k/\epsilon$ [s], and a turbulent viscosity can be formulated as $v_T = Ck^2/\epsilon$ $[\text{m}^2\text{s}^{-1}]$. Since all quantities can be formed having these variables, the two-equation models are *complete models*. For instance, flow-dependent specifications such as $\ell_m(\mathbf{x}, t)$ [m] are not required.

The model transport equation for k $[\text{m}^2\text{s}^{-2}]$ is already provided in the previous section. The specification of turbulent viscosity is provided as

$$v_T = C_\mu k^2/\epsilon \, , \tag{15.26}$$

where $C_\mu = 0.09$ $[-]$ is one of five model constants. This relationship implies that v_T $[\text{m}^2\text{s}^{-1}]$ only depends on k $[\text{m}^2\text{s}^{-2}]$ and ϵ $[\text{m}^2\text{s}^{-3}]$, and not gradients of mean velocity.

Unlike the model equation for k $[\text{m}^2\text{s}^{-2}]$ that can be derived, the model equation for ϵ $[\text{m}^2\text{s}^{-3}]$ is mainly empirical, and in some authors' opinion a pure invention [8]

$$\underbrace{\frac{\overline{D\epsilon}}{\overline{Dt}}}_{\text{Mean Substantial Derivative}} \equiv \underbrace{\frac{\partial \epsilon}{\partial t}}_{\text{Storage}} + \underbrace{\langle \mathbf{U} \rangle . \nabla \epsilon}_{\text{Advection}}$$

$$= \underbrace{\nabla . \left(\frac{v_T}{\sigma_\epsilon} \nabla \epsilon \right)}_{\text{Dissipation Flux Divergence}} + \underbrace{C_{\epsilon 1} \frac{\mathcal{P}\epsilon}{k}}_{\text{Production}} - \underbrace{C_{\epsilon 2} \frac{\epsilon^2}{k}}_{\text{Dissipation}} \, . \tag{15.27}$$

There are many variations of the $k - \epsilon$ model. The constants for the *standard model* are provided as [7]

$$C_\mu = 0.09, C_{\epsilon 1} = 1.44, C_{\epsilon 2} = 1.92, \sigma_k = 1.0, \sigma_\epsilon = 1.3 . \tag{15.28}$$

15.6   The $k - \omega$ Model

Two-equation models are numerous and expand beyond the $k - \epsilon$ model. In many of such models k [m^{2}s^{-2}] is taken as one of the variables; however, there are diverse options for the second variable. The $k - \omega$ model was first proposed by [5], where the second variable was defined as $\omega \equiv \epsilon/k$ [s^{-1}]. It has been suggested that the choice of this second variable mainly impacts simulations of non-homogeneous turbulence [4] because for homogeneous turbulence the equation for the second variable reduces to the ϵ [m^{2}s^{-3}] equation provided earlier. Therefore, the careful choice of this second variable may help developing better models for specific non-homogeneous flows.

For non-homogeneous flows, the difference usually lies in the flux divergence term. The model equation for ω [s^{-1}] can be given as

$$\underbrace{\frac{D\omega}{Dt}}_{\text{Mean Substantial Derivative}} \equiv \underbrace{\frac{\partial\omega}{\partial t}}_{\text{Storage}} + \underbrace{\langle U \rangle . \nabla\omega}_{\text{Advection}}$$

$$= \underbrace{\nabla . \left(\frac{\nu_T}{\sigma_\omega} \nabla\omega \right)}_{\text{Dissipation Flux Divergence}} + \underbrace{C_{\omega 1} \frac{\mathcal{P}\omega}{k}}_{\text{Production}} - \underbrace{C_{\omega 2} \omega^2}_{\text{Dissipation}} . \tag{15.29}$$

If one tries to derive this equation from the ϵ equation in the $k - \epsilon$ model, the difference between the models can be observed. Assuming $\sigma_k = \sigma_\epsilon = \sigma_\omega$ [-], the ω equation can be derived as

$$\frac{D\omega}{Dt} = \nabla . \left(\frac{\nu_T}{\sigma_\omega} \nabla\omega \right) + (C_{\epsilon 1} - 1) \frac{\mathcal{P}\omega}{k} - (C_{\epsilon 2} - 1)\omega^2 + \frac{2\nu_T}{\sigma_\omega k} \nabla\omega . \nabla k . \tag{15.30}$$

For homogeneous turbulence the ω and ϵ equations are the same, and the choices of model constants are $C_{\omega 1} = C_{\epsilon 1} - 1$ [-] and $C_{\omega 2} = C_{\epsilon 2} - 1$ [-]. However, for non-homogeneous flows, the two equations will be different since there will be an extra term in the ϵ equation for the $k - \epsilon$ model written as a $k - \omega$ model.

Reference [9] successfully combined functionalities of the $k - \epsilon$ and $k - \omega$ models such that the two are blended. In the free stream, and away from the walls, the blended model behaves as a $k - \epsilon$ model, which is known to perform better in homogeneous turbulence; however, near the walls, the blended model behaves as a $k - \omega$ model with superior performance for non-homogeneous turbulence.

15.7 Turbulent-Viscosity Models for the Atmospheric Boundary Layer

The Atmospheric Boundary Layer (ABL) is the fraction of the atmosphere near the earth surface that responds to surface forcing within a timescale of less than about one hour [10]. Most operational weather models parameterize turbulence in ABL in the form of vertical and/or horizontal diffusion.

In ABL flows, it is typically assumed that the coordinate direction normal to the earth surface is z (not y). We first introduce the concept of potential temperature. Since the atmosphere thins in density and pressure with increasing altitude, instead of the real temperature the potential temperature is used, which is defined as

$$\langle \Theta \rangle = \langle T \rangle \left(\frac{P_0}{\langle P \rangle} \right)^{\frac{R}{c_p}} , \qquad (15.31)$$

where $\langle \Theta \rangle$ [K] is the equivalent temperature (or potential temperature) for a parcel of air at temperature $\langle T \rangle$ [K] and pressure $\langle P \rangle$ [Pa], which is adiabatically moved to a reference pressure P_0 [Pa] [11]. Here R [JK^{-1}mol^{-1}] is the gas constant of air, and c_p [JK^{-1}mol^{-1}] is the specific heat capacity of air at a constant pressure. In meteorology it is assumed that $\frac{R}{c_p} = 0.286$ [−] for air. Note that Reynolds decomposition is still used for the potential temperature, where $\Theta = \langle \Theta \rangle + \theta$ [K]. In meteorology, sometimes the virtual potential temperature $\Theta_v = \langle \Theta_v \rangle + \theta_v$ [K] is also used. Virtual potential temperature is the theoretical potential temperature of dry air that would have the same density as moist air. Using Reynolds averaging, the ABL equations of continuity, momentum, and heat can be listed as [10, 12]

$$\nabla . \langle \mathbf{U} \rangle = 0 \qquad (15.32)$$

$$\frac{\overline{D} \langle U \rangle}{\overline{D} t} \equiv \frac{\partial \langle U \rangle}{\partial t} + \langle \mathbf{U} \rangle . \nabla \langle U \rangle = f (\overline{V} - \overline{V}_g) - \frac{\langle u u_j \rangle}{\partial x_j} \qquad (15.33)$$

$$\frac{\overline{D} \langle V \rangle}{\overline{D} t} \equiv \frac{\partial \langle V \rangle}{\partial t} + \langle \mathbf{U} \rangle . \nabla \langle V \rangle = f (\overline{U}_g - \overline{U}) - \frac{\langle v u_j \rangle}{\partial x_j} \qquad (15.34)$$

$$\frac{\overline{D} \langle \Theta_v \rangle}{\overline{D} t} \equiv \frac{\partial \langle \Theta_v \rangle}{\partial t} + \langle \mathbf{U} \rangle . \nabla \langle \Theta_v \rangle = - \frac{1}{\langle \rho \rangle C_p} \left(L_v \frac{\partial \langle E_j \rangle}{\partial x_j} + \frac{\partial \langle Q_j^* \rangle}{\partial x_j} \right) - \frac{\langle \theta_v u_j \rangle}{\partial x_j} , \qquad (15.35)$$

where $\langle U \rangle$ and $\langle V \rangle$ [ms^{-1}] are mean wind velocity components along horizontal x and y directions, respectively. Note that the mean wind velocity component in the vertical z direction, i.e. $\langle W \rangle$ [ms^{-1}], is assumed zero. f [s^{-1}] is the Coriolis parameter, subscript g denotes geostrophic wind, $\langle \rho \rangle$ [kgm^{-3}] is the mean density, C_p [Jkg^{-1}K^{-1}] is the heat capacity of air at constant pressure, L_v [Jkg^{-1}] is the latent heat of vaporization, E_j [kgm^{-2}s^{-1}] is the jth component of the water vapour flux, and Q_j^* [Wm^{-2}] is the jth component of the net radiation flux.

The parameterization of turbulence in ABL involves determination of the terms $\langle u_i u_j \rangle$ [m^{2}s^{-2}] and $\langle \theta_v u_j \rangle$ [Kms^{-1}], which often involve either separate transport equations for each term (or a combination of terms) or further closure schemes that require other model constants [13]. While the continuity, momentum, and energy equations are common among many ABL turbulence models, there are a variety of models that parameterize the turbulent terms [14–19]. It is beyond the scope of this text to discuss the details of the variety of ABL turbulence models; however, generally simple models employ fewer transport equations and model constants, while complex models employ more transport equations and model constants [13]. A common theme for many ABL turbulence models involves parameterization of the transport equation for turbulence kinetic energy k [m^{2}s^{-2}] and closure of a mixing length ℓ [m] as a model constant. In addition, many models parameterize the turbulent terms $\langle u_i u_j \rangle$ [m^{2}s^{-2}] and $\langle \theta_v u_j \rangle$ [Kms^{-1}] using either the down-gradient or counter-gradient hypotheses [12].

Problems

15.1 Explain why the algebraic and turbulence kinetic energy models are incomplete turbulence models, while the $k - \epsilon$ model is a complete turbulence model.

15.2 Consider the log-law region of a wall-bounded flow. Suppose that in the log-law region the turbulence kinetic energy and the friction velocity can be related using $u_\tau = C_\mu^{1/4} k^{1/2}$ [ms^{-1}], where $C_\mu = 0.09$ [−]. Also consider that in the log-law region the friction velocity is dominated by the Reynolds stress, i.e. $-\langle uv \rangle = u_\tau^2$ [m^{2}s^{-2}]. Use the log-law and the specification $\ell_m = \kappa y$ [m] to show that the appropriate value of the constant c [−] in the relation $\nu_T = c k^{1/2} \ell_m$ [m^{2}s^{-1}] is

$$c \approx 0.55 . \tag{15.36}$$

15.3 An atmospheric scientist wishes to develop a transient one-dimensional (1D) momentum transport model. She closes her turbulence model by parameterizing a mixing length that also formulates the turbulent viscosity. Suppose she uses the Cartesian coordinate system with coordinate axes of x, y, and z, and velocities corresponding to these axes being $U = \langle U \rangle + u$, $V = \langle V \rangle + v$, and $W = \langle W \rangle + w$ [ms^{-1}], respectively. Further, she assumes that mean flow is only in the x direction parallel to the surface and that the direction z is normal to the surface. She assumes mean velocities in y and z directions being zero, i.e. $\langle V \rangle = \langle W \rangle = 0$ ms^{-1}. In addition, she assumes that the mean velocity $\langle U \rangle$ [ms^{-1}] in the x and y directions does not change. She also assumes that the modified pressure has a constant gradient in the x direction. Her 1D transport model is written as

$$\frac{\partial \langle U \rangle}{\partial t} = \frac{\partial}{\partial z} \left(\nu_T \frac{\partial \langle U \rangle}{\partial z} \right) - \tau , \tag{15.37}$$

$$\begin{cases} \nu_T = \ell_m^2 \left| \frac{d\langle U \rangle}{dz} \right| , \\ \ell_m = \kappa z / \left(1 + \frac{\kappa z}{\ell_0} \right) . \end{cases}$$

In the momentum equation identify the following terms: storage, surface forces and Reynolds stress, and modified pressure forces. In the above equations, identify the parameterizations for the following terms: turbulent viscosity and mixing length. Demonstrate why the advection terms in the momentum transport equation have vanished. Which classification of turbulence model is this simple model according to this chapter?

15.4 An atmospheric scientist wishes to develop a transient one-dimensional (1D) momentum and turbulence kinetic energy transport model. She closes her turbulence model by assuming turbulent Prandtl number $\sigma_k = 1$ [−] and parameterizing a mixing length that also formulates the turbulent viscosity and the turbulence kinetic energy dissipation rate. Suppose she uses the Cartesian coordinate system with coordinate axes of x, y, and z, and velocities corresponding to these axes being $U = \langle U \rangle + u$, $V = \langle V \rangle + v$, and $W = \langle W \rangle + w$ [ms^{-1}], respectively. Further, she assumes that mean flow is only in the x direction parallel to the surface and that the direction z is normal to the surface. She assumes mean velocities in y and z directions being zero, i.e. $\langle V \rangle = \langle W \rangle = 0$ ms^{-1}. In addition, she assumes that the mean velocity $\langle U \rangle$ [ms^{-1}] in the x and y directions does not change. She also assumes that the modified pressure has a constant gradient in the x direction. Her 1D transport model is written as

$$\frac{\partial \langle U \rangle}{\partial t} = \frac{\partial}{\partial z} \left(\nu_T \frac{\partial \langle U \rangle}{\partial z} \right) - \tau , \tag{15.38}$$

$$\frac{\partial k}{\partial t} = \frac{\partial}{\partial z} \left(\frac{\nu_T}{\sigma_k} \frac{\partial k}{\partial z} \right) + \nu_T \left(\frac{\partial \langle U \rangle}{\partial z} \right)^2 - \epsilon , \tag{15.39}$$

$$\begin{cases} \nu_T = C_k \ell_m k^{1/2}, \\ \epsilon = C_\epsilon \ell_m^{-1} k^{3/2}, \\ \ell_m = \kappa z / \left(1 + \frac{\kappa z}{\ell_0} \right) . \end{cases}$$

In the above, identify the momentum and turbulence kinetic energy equations. In the momentum equation identify the following terms: storage, surface forces and Reynolds stress, and modified pressure forces. In the turbulence kinetic energy equation identify the following terms: storage, energy flux divergence, shear production, and turbulence kinetic energy dissipation rate. In the above equations, identify the parameterizations for the following terms: turbulent viscosity, turbulence kinetic energy dissipation rate, and mixing length. Demonstrate why the advection terms in the momentum and turbulence kinetic energy transport equations have vanished. Which classification of turbulence model is this simple model according to this chapter?

15.5 An atmospheric scientist wishes to develop a steady one-dimensional (1D) heat, momentum, and turbulence kinetic energy transport model. He assumes a non-constant turbulent viscosity ν_T [m^{2}s^{-1}] that accounts for effects of molecular and turbulent diffusion and a turbulent Prandtl number $Pr_T = \sigma_k = 1$ [−]. He assumes a Cartesian coordinate system with coordinate axes of x, y, and z, and velocities corresponding to these axes being $U = \langle U \rangle + u$, $V = \langle V \rangle + v$, and $W = \langle W \rangle + w$ [ms^{-1}], respectively. Further, he assumes that mean flow is only in the x direction parallel to the surface and that the direction z is normal to the surface, i.e. $\langle V \rangle = \langle W \rangle = 0$ ms^{-1}. In addition, he assumes that the mean velocity $\langle U \rangle$ [ms^{-1}] in the x and y directions does not change. He assumes a constant heat sink or source for temperature by a uniform rate of cooling or heating in the domain. He also assumes that the modified pressure has a constant gradient in the x direction. His 1D transport model is written as

$$0 = \frac{\partial}{\partial z} \left(\frac{\nu_T}{Pr_T} \frac{\partial \langle T \rangle}{\partial z} \right) - \gamma \ , \tag{15.40}$$

$$0 = \frac{\partial}{\partial z} \left(\nu_T \frac{\partial \langle U \rangle}{\partial z} \right) - \tau \ , \tag{15.41}$$

$$0 = \frac{\partial}{\partial z} \left(\frac{\nu_T}{\sigma_k} \frac{\partial k}{\partial z} \right) + \nu_T \left(\frac{\partial \langle U \rangle}{\partial z} \right)^2 - \frac{g}{T_0} \frac{\nu_T}{Pr_T} \frac{\partial \langle T \rangle}{\partial z} - \epsilon \ , \tag{15.42}$$

$$\begin{cases} \nu_T = C_k \ell_m k^{1/2}, \\ \epsilon = C_\epsilon \ell_m^{-1} k^{3/2}, \\ \ell_m = \kappa z / \left(1 + \frac{\kappa z}{\ell_0} \right) . \end{cases}$$

In the above, identify the heat, momentum, and turbulence kinetic energy equations. In the heat equation, identify the following terms: storage, diffusion of mean temperature, and rate of heat sink or source. In the momentum equation identify the following terms: storage, surface forces and Reynolds stress, and modified pressure forces. In the turbulence kinetic energy equation identify the following terms: storage, energy flux divergence, shear production, buoyant production or sink, and turbulence kinetic energy dissipation rate. In the above equations, identify the parameterizations for the following terms: turbulent viscosity, turbulence kinetic energy dissipation rate, and mixing length. Demonstrate why the advection terms in the heat, momentum, and turbulence kinetic energy transport equations have vanished. Which classification of turbulence model is this simple model according to this chapter?

15.6 An atmospheric scientist wishes to develop a transient one-dimensional (1D) passive scalar, momentum, and turbulence kinetic energy transport model. He assumes a non-constant turbulent viscosity v_T [m²s⁻¹] that accounts for effects of molecular and turbulent diffusion, a turbulent Prandtl number $\sigma_k = 1$ [−], and a turbulent Schmidt number $Sc_T = 1$ [−]. He assumes a Cartesian coordinate system with coordinate axes of x, y, and z, and velocities corresponding to these axes being $U = \langle U \rangle + u$, $V = \langle V \rangle + v$, and $W = \langle W \rangle + w$ [ms⁻¹], respectively. Further, he assumes that mean flow is only in the x direction parallel to the surface and that the direction z is normal to the surface, i.e. $\langle V \rangle = \langle W \rangle = 0$ ms⁻¹. In addition, he assumes that the mean velocity $\langle U \rangle$ [ms⁻¹] in the x and y directions does not change. He also assumes that the modified pressure has a constant gradient in the x direction. His 1D transport model is written as

$$\frac{\partial \langle \phi \rangle}{\partial t} = \frac{\partial}{\partial z} \left(\frac{v_T}{Sc_T} \frac{\partial \langle \phi \rangle}{\partial z} \right) , \tag{15.43}$$

$$\frac{\partial \langle U \rangle}{\partial t} = \frac{\partial}{\partial z} \left(v_T \frac{\partial \langle U \rangle}{\partial z} \right) - \tau , \tag{15.44}$$

$$\frac{\partial k}{\partial t} = \frac{\partial}{\partial z} \left(\frac{v_T}{\sigma_k} \frac{\partial k}{\partial z} \right) + v_T \left(\frac{\partial \langle U \rangle}{\partial z} \right)^2 - \epsilon , \tag{15.45}$$

$$\begin{cases} v_T = C_k \ell_m k^{1/2}, \\ \epsilon = C_\epsilon \ell_m^{-1} k^{3/2}, \\ \ell_m = \kappa z / \left(1 + \frac{\kappa z}{\ell_0} \right) . \end{cases}$$

In the above, identify the passive scalar, momentum, and turbulence kinetic energy equations. In the passive scalar equation, identify the following terms: storage and diffusion of mean passive scalar. In the momentum equation identify the following terms: storage, surface forces and Reynolds stress, and modified pressure forces. In the turbulence kinetic energy equation identify the following terms: storage, energy flux divergence, shear production, and turbulence kinetic energy dissipation rate. In the above equations, identify the parameterizations for the following terms: turbulent viscosity, turbulence kinetic energy dissipation rate, and mixing length. Demonstrate why the advection terms in the passive scalar, momentum, and turbulence kinetic energy transport equations have vanished. Which classification of turbulence model is this simple model according to this chapter?

15.7 Explain what a mixing length physically describes in turbulence modelling.

15.8 In wall flows, the mixing length algebraic turbulence model provides the turbulent viscosity as

$$v_T = \ell_m^2 \left| \frac{\partial \langle U \rangle}{\partial y} \right| , \qquad (15.46)$$

where ℓ_m [m] is the mixing length, $\langle U \rangle$ [ms^{-1}] is mean flow parallel to the wall surface, and y is the wall-normal direction. In the log-law region, the mixing length can be given by $\ell_m = \kappa y$ [m], where κ [−] is the von Kármán constant. Show that the turbulent viscosity in the log-law region can be given by:

$$v_T = u_\tau \kappa y , \qquad (15.47)$$

where u_τ [ms^{-1}] is the friction velocity.

References

1. Sonntag R E, Van Wylen G J (1966) Fundamentals of statistical thermodynamics. Wiley, New York
2. von Kármán T (1931) Mechanical similitude and turbulence. National Advisory Committee for Aeronautics, Washington DC
3. Spalart P R, Allmaras S R (1994) A one-equation turbulence model for aerodynamic flows. Recherche Aérospatiale 1:5–21
4. Pope S B (2000) Turbulent flows. Cambridge University Press, Cambridge
5. Kolmogorov A N (1942) The equations of turbulent motion in an incompressible fluid. Izvestia Acad Sci USSR Phys 6:56–58
6. Prandtl L (1945) Über ein neues formelsystem für die ausgebildete turbulenz. Nachr Akad Wiss Göttingen Math-Phys K1:6–19
7. Launder B E, Spalding D B (1974) The numerical computation of turbulent flows. Comp Meth Appl Mech Eng 3:269–289
8. Davidson P A (2005) Turbulence: an introduction for scientists and engineers. Oxford University Press, Oxford
9. Menter F (1994) Two-equation eddy-viscosity turbulence models for engineering applications. AIAA J 32:1598–1605
10. Stull R B (1988) An introduction to boundary layer meteorology. Kluwer Academic Publishers, Dordrecht
11. Aliabadi A A, Staebler R M, de Grandpré J et al (2016) Comparison of estimated atmospheric boundary layer mixing height in the Arctic and southern Great Plains under statically stable conditions: experimental and numerical aspects. Atmos-Ocean 54:60–74
12. Aliabadi A A, Staebler R M, Liu M et al (2016) Characterization and parametrization of Reynolds stress and turbulent heat flux in the stably-stratified lower Arctic troposphere using aircraft measurements. Bound-Lay Meteorol 161:99–126
13. Mellor G L, Yamada T (1974) A hierarchy of turbulence closure models for planetary boundary layers. J Atmos Sci 31:1791–1806
14. Janjić Z I (1994) The step-mountain eta coordinate model: further developments of the convection, viscous sublayer, and turbulence closure schemes. Mon Weather Rev 122:927–945
15. Janjic Z I, Gerrity Jr J P, Nickovic S (2001) An alternative approach to nonhydrostatic modeling. Mon Weather Rev 129:1164–1178

16. Hong S-Y, Noh Y, Dudhia J (2006) A new vertical diffusion package with an explicit treatment of entrainment processes. Mon Weather Rev 134:2318–2341
17. Gibbs J A, Fedorovich E, van Eijk A M J (2011) Evaluating Weather Research and Forecasting (WRF) model predictions of turbulent flow parameters in a dry convective boundary layer. J Appl Meteorol Clim 50:2429–2444
18. Xue L, Chu X, Rasmussen R et al (2014) The dispersion of silver iodide particles from ground-based generators over complex terrain. Part II: WRF large-eddy simulations versus observations. J Appl Meteorol Clim 53:1342–1361
19. Nambiar M K, Robe F, Seguin A M et al (2020) Diurnal and seasonal variation of area-fugitive methane advective flux from an open-pit mining facility in northern Canada using WRF. Atmosphere-Basel 11:1227

Chapter 16
Large-Eddy Simulation Models

Abstract This chapter introduces the Large-Eddy Simulation (LES) technique. The concept of spatial filtering of flow properties is discussed. The residual or Subgrid-Scale (SGS) component of the velocity vector field is defined. The continuity and momentum equations are developed for spatially filtered velocity vector field. SGS models are introduced as closure parameterizations that describe the modelled, or non-resolved, scales of motion. Two SGS models are provided: the Smagorinsky model and the one-equation turbulence kinetic energy model. The problem of inlet condition for LES is addressed, involving a variety of techniques to produce flow fluctuations at the inlet of an LES domain. Finally, a synthetic inlet turbulence generator for atmospheric boundary layers is provided.

16.1 Preliminaries

In Large-Eddy Simulation (LES), the larger three-dimensional unsteady turbulent motions are directly represented, while the effects of the smaller-scale motions are modelled. In terms of computational expense, LES lies between Reynolds-stress models and Direction Numerical Simulation (DNS). It is motivated by the limitations of each of these approaches. In addition, for many applications, most dominant transport mechanisms occur at larger scales (e.g. pollution and heat transport), hence justifying simulating large scales accurately while modelling small scales. From theoretical point of view, large scales of turbulent motion are dependent on geometry and conditions of the flow, while smaller scales can be universal. This in turn motivates the use of LES for more accurate predictions. LES was pioneered for applications in meteorology first in the 1960s and 1970s [1, 2]. It is continually being used in meteorological models [3, 4].

LES requires four conceptual steps. First, a *filtering operation* is required to decompose the velocity $\mathbf{U}(\mathbf{x}, t)$ [ms^{-1}] into the sum of a filtered (or resolved) component $\overline{\mathbf{U}}(\mathbf{x}, t)$ [ms^{-1}] and a *residual* or *Subgrid-Scale (SGS)* component $\mathbf{u}'(\mathbf{x}, t)$ [ms^{-1}]. The filtered velocity field differs from a mean $\langle \mathbf{U}(\mathbf{x}, t) \rangle$ [ms^{-1}] because it is time dependent and represents the motion of the large eddies:

$$\mathbf{U}(\mathbf{x}, t) = \overline{\mathbf{U}}(\mathbf{x}, t) + \mathbf{u}'(\mathbf{x}, t) \ . \tag{16.1}$$

In fact, $\overline{\mathbf{U}}(\mathbf{x}, t)$ [ms^{-1}] is a form of a spatial average, introduced earlier using Eq. 3.42. Second, the Navier–Stokes momentum Eq. 2.8 can be developed for the evolution of the filtered velocity. This equation will contain a *residual stress tensor* or *SGS tensor* that arises from the residual motions. Third, the Navier–Stokes momentum equation must be closed by modelling the residual-stress tensor or SGS tenser, usually using an eddy-viscosity (or down-gradient diffusion) hypothesis introduced earlier using Eq. 4.22. Fourth, the model equations are solved numerically for $\overline{\mathbf{U}}(\mathbf{x}, t)$ [ms^{-1}], which provides an approximation of the large-scale motions in one realization of the turbulent flow. It must be noted that LES always provides transient solutions for the flow variables.

16.2   Filtering

Mathematically, the spatial filtering operation is described by integrating a function over its entire domain using a filter function. Figure 16.1 shows the spatial filtering process for a bottle using a filter function with length Δ [m]. After filtering, all variations of the function with a lengthscale smaller than Δ [m] are smoothed.

For turbulence, the function can be the velocity vector, the integrand, and can be integrated over the entire flow domain

$$\overline{\mathbf{U}}(\mathbf{x}, t) = \int G(\mathbf{r}, \mathbf{x})\mathbf{U}(\mathbf{x} - \mathbf{r}, t)d\mathbf{r} \ , \tag{16.2}$$

where the specified filter function G [m^{-1}] must satisfy the normalization condition:

$$\int G(\mathbf{r}, \mathbf{x})d\mathbf{r} = 1 \ . \tag{16.3}$$

Fig. 16.1 Spatial filtering of a bottle using a filter function with length Δ [m]

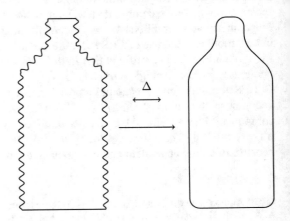

The filtering operation can be demonstrated for a 1D velocity field $\overline{U}(x)$ [ms^{-1}] with a homogeneous filter. A commonly used filter is the box filter. With this filter, $\overline{U}(x)$ [ms^{-1}] is simply the average of $U(x')$ [ms^{-1}] in the interval $x - \frac{1}{2}\Delta < x' < x + \frac{1}{2}\Delta$ [m], where Δ [m] can be understood as the filter width. The box filter $G(r, x)$ [m^{-1}] can be given as

$$\begin{cases} \frac{1}{\Delta} & \text{if } |x - r| < \frac{1}{2}\Delta , \\ 0 & \text{otherwise} . \end{cases}$$

16.3   Filtered Conservation Equations

By applying the filtering operation to the Navier–Stokes equations we can obtain the governing equations for filtered quantities. When spatially uniform filters are used, such as the box filter, the filtering and differentiation operations commute. The filtered continuity Eq. 2.2 is

$$\overline{\left(\frac{\partial U_i}{\partial x_i} \right)} = \frac{\partial \overline{U}_i}{\partial x_i} = 0 . \tag{16.4}$$

This also implies a continuity equation for the SGS velocity. Therefore, both the filtered field $\overline{\mathbf{U}}$ [ms^{-1}] and the SGS field $\mathbf{u}'$ [ms^{-1}] are solenoidal, i.e.

$$\frac{\partial u_i'}{\partial x_i} = \frac{\partial}{\partial x_i}(U_i - \overline{U}_i) = 0 . \tag{16.5}$$

The filtered momentum conservation Eq. 2.8 can be expressed as

$$\underbrace{\frac{\partial \overline{U}_j}{\partial t}}_{\text{Storage}} + \underbrace{\frac{\partial \overline{U_i U_j}}{\partial x_i}}_{\text{Advection}} = \underbrace{v \frac{\partial^2 \overline{U}_j}{\partial x_i \partial x_i}}_{\text{Surface Forces}} - \underbrace{\frac{1}{\rho} \frac{\partial \overline{p}}{\partial x_j}}_{\text{Normal and Body Forces}} , \tag{16.6}$$

where $\overline{p}(\mathbf{x}, t)$ [kgm^{-1}s^{-2}] is the filtered pressure field. This equation needs further attention since the filtered product $\overline{U_i U_j}$ [m^{2}s^{-2}] is different than the product of the filtered velocities $\overline{U}_i \overline{U}_j$ [m^{2}s^{-2}]. The difference of the two is the *residual stress tensor* defined by:

$$\tau_{ij}^R \equiv \overline{U_i U_j} - \overline{U}_i \overline{U}_j . \tag{16.7}$$

This is analogous to the *Reynolds stress tensor* $\langle u_i u_j \rangle \equiv \langle U_i U_j \rangle - \langle U_i \rangle \langle U_j \rangle$ [m^{2}s^{-2}]. The *residual kinetic energy* is defined as

$$k_r \equiv \tau_{ii}^R . \tag{16.8}$$

The *anisotropic residual stress tensor* is defined, analogous to Eq. 4.12, by:

$$\tau_{ij}^r = \tau_{ij}^R - \frac{2}{3} k_r \delta_{ij} . \tag{16.9}$$

It is possible to absorb the isotropic residual stress in the pressure to obtain the *modified filtered pressure* such that

$$\overline{p}^m \equiv \overline{p} + \frac{2}{3} \rho k_r . \tag{16.10}$$

With these considerations it is possible to obtain the momentum Eq. 2.8 for the filtered velocity as a function of anisotropic residual-stress tensor such that

$$\underbrace{\frac{\overline{D}\,\overline{U}_j}{\overline{Dt}}}_{\text{Mean Substantial Derivative}} = \underbrace{\nu \frac{\partial^2 \overline{U}_j}{\partial x_i \partial x_i}}_{\text{Surface Forces}} - \underbrace{\frac{\partial \tau_{ij}^r}{\partial x_i}}_{\text{Anisotropic Residual} - \text{Stress Forces}}$$

$$- \underbrace{\frac{1}{\rho} \frac{\partial \overline{p}^m}{\partial x_j}}_{\text{Modified Pressure Forces}} . \tag{16.11}$$

The material or substantial derivative based on the filtered velocity can be defined in a similar way to the mean velocity given in Eq. 4.8, i.e.

$$\frac{\overline{D}}{\overline{Dt}} \equiv \frac{\partial}{\partial t} + \overline{\mathbf{U}} . \nabla . \tag{16.12}$$

The closure for the filtered momentum equation can be obtained by modelling the anisotropic residual or SGS stress tensor. For brevity, sometimes this stress tensor is referred to as residual or SGS stress tensor, with the word anisotropic dropped.

16.4   The Smagorinsky Model

The simplest model to close the turbulence model, i.e. Eq. 16.9, for LES was proposed by [1]. First, the model employs the linear eddy-viscosity relationship to give the anisotropic residual-stress tensor

$$\tau_{ij}^r = -2\nu_r \overline{S}_{ij} , \tag{16.13}$$

where $\overline{S}_{ij}$ [s^{-1}] is the filtered rate of strain, as defined by Eq. 2.17 but for the spatially filtered velocity field. Second, the model specifies the residual eddy viscosity $v_r(\mathbf{x}, t)$ [m^{2}s^{-1}] by employing a mixing length relationship, analogous to Eq. 15.12, such that

$$v_r = \ell_S^2 \overline{S} = (C_S \Delta)^2 \overline{S} , \tag{16.14}$$

where $\overline{S}$ [s^{-1}] is the *characteristic filtered rate-of-strain*. ℓ_S [m] is *Smagorinsky lengthscale*, analogous to the mixing length, which through constant C_S [–] can be related to the filter width Δ [m]. In a simplistic model it can be assumed $C_S = 0.09$ [–]. In many LES numerical simulation models Δ [m] is taken as the geometric average of the grid element dimensions [5, 6]

$$\Delta = (\Delta x \Delta y \Delta z)^{1/3} . \tag{16.15}$$

16.5 One-equation Turbulence Kinetic Energy Model

Another recent approach to close the turbulence model for LES is the one-equation turbulence kinetic energy model [5, 7, 8]. In this model, an extra transport equation is solved for the SGS Turbulence Kinetic Energy (TKE) k_{sgs} [m^{2}s^{-2}] given as

$$\underbrace{\frac{\overline{D}k_{sgs}}{\overline{D}t}}_{\text{Mean Substantial Derivative}} \equiv \underbrace{\frac{\partial k_{sgs}}{\partial t}}_{\text{Storage}} + \underbrace{\overline{\mathbf{U}}.\nabla k_{sgs}}_{\text{Advection}}$$

$$= \underbrace{\nabla.\left(\frac{v_r}{\sigma_{k_{sgs}}}\nabla k_{sgs}\right)}_{\text{Energy Flux Divergence}} + \underbrace{\mathcal{P}}_{\text{Production}} - \underbrace{\epsilon}_{\text{Dissipation}} . \tag{16.16}$$

Here, the SGS Turbulence Kinetic Energy (TKE) k_{sgs} [m^{2}s^{-2}] does not represent fluctuations at all scales but only at smallest scales below the filter width Δ [m]. $\overline{\mathbf{U}}$ [ms^{-1}] is the spatially and temporally resolved velocity vector and $\sigma_{k_{sgs}} = 1$ [–] is the turbulent Prandtl number for SGS TKE. The production term $\mathcal{P}$ [m^{2}s^{-3}] is closed as

$$\mathcal{P} = -\tau_{ij}^r \overline{S}_{ij} = 2v_r \left(\overline{S}_{ij}\right)^2 , \tag{16.17}$$

where the residual viscosity v_r [m^{2}s^{-1}] is closed using the following relationship employing the SGS TKE k_{sgs} [m^{2}s^{-2}] and more constants

$$v_r = C_k k_{sgs}^{1/2} l . \tag{16.18}$$

Here $C_k = 0.094$ [–] is a constant, and $l = C_\Delta (\Delta x \Delta y \Delta z)^{1/3}$ [m] is the subgrid mixing length. Constant C_Δ [–] is usually in the order of one but can be optimized to control the SGS dissipation rate of turbulence kinetic energy [5]. Finally the dissipation rate of turbulence kinetic energy ϵ [m^{2}s^{-3}] is closed using one more constant

$$\epsilon = C_\epsilon \frac{k_{sgs}^{3/2}}{l} , \tag{16.19}$$

with $C_\epsilon = 1.048$ [–]. The one-equation turbulence kinetic energy model for closing LES is advantageous over the Smagorinsky model, especially for free shear flows where gradients of velocity may approach zero in the domain while the residual viscosity is not negligible. This is analogous to the advantage of turbulence kinetic energy models over mixing length models for turbulent viscosity models, as discussed in Chap. 15.

16.6   The Problem of Inlet Condition

As appealing as LES may appear, it has a serious limitation: realistic inlet boundary condition. In order to build a robust LES model, introducing realistic turbulent fluctuations at the inlet are required that would evolve in the entire domain. From a theoretical stand point, the fluctuations must meet several criteria: (a) they must be stochastically varying, on scales down to the spatial and temporal filter scales; (b) they must be compatible with the Navier–Stokes equations; (c) they must be composed of coherent eddies across a range of spatial scales down to the filter length; (d) they must allow easy specification of turbulence properties; and (e) they must be easy to implement [9].

Two of the most common approaches to generate the inlet turbulent fluctuations for LES models are the synthetic and precursor methods. In the synthetic method, random fields are constructed at the inlet, while in the precursor method an additional simulation is performed to generate the desired fluctuations. Precursor methods are shown to be more accurate but more computationally demanding and more difficult to implement [9]. These methods have been reviewed in the literature [9, 10]. The synthetic method is more popular for practical applications and is described further below.

Lund et al. [11] developed a synthetic model, originally introduced by [12], to generate the inlet turbulent fluctuations by rescaling the velocity field at a downstream station, re-introducing it as a boundary condition at the inlet, and hence developing spatial and temporal turbulent boundary layers economically [11, 13]. Compared to primitive methods of random inclusion of perturbations at inlet, it has been shown that this synthetic method reduces adaptation distance upstream of the flow significantly, resulting in smaller domains and more economical simulations [11]. Another common synthetic model is the vortex method originally developed by [14] and later refined by [15, 16], and [17] that inserts random two-dimensional vortices at the inlet boundary that evolve into the simulation domain. These vortices are parameterized by realistic lengthscales, timescales, and vorticity magnitudes, formulated from mean flow information and grid spacing. [18] developed a method based on synthesizing random divergence-free turbulence velocities with consideration of spectra and coherency functions that match the flow statistics. The method is also known as Consistent Discrete Random Field Generation (CDRFG). This scheme maintains both the turbulence spectra and coherency function, which are essential for proper simulation of interaction of turbulent flow with flexible structures, such as buildings, prone to flow-induced dynamic excitation. Another notable synthetic method useful for analysis of flow-induced excitation of structures is known as Modified Discretizing and Synthesizing Random Flow Generation (MDSRFG) that was developed recently by [19] and [20]. This method is based on providing a true representation of the coherency of the velocity field at the inlet. An alternative technique to the classical velocity perturbations is the Temperature Perturbation Method (TPM) developed by [21] to generate inlet turbulence fluctuations. This method relies on the creation of turbulent structures through a buoyancy triggered mechanism by seeding the flow with random temperature perturbations at the inlet. Buckingham et al. [21] found that the temperature perturbation method can result in long adaptation distances. Although this method benefits from the simplicity of not requiring prior knowledge of second order moments or integral lengthscales at the inlet, additional refinements for developing flows that include physical temperature effects are required.

16.7   A Synthetic Inlet Turbulence Generator for Atmospheric Boundary Layers

To generate turbulence at the inlet of models for atmospheric boundary layers, a vortex method may be used. A vortex method was developed by [5] and [6] and is described here. The original version was developed by [14] and has been continually improved until recently [17]. The main idea of the vortex method is to generate velocity fluctuations in the form of synthetic eddies derived from mean statistical information about the flow as a function of space (height above ground) and time. To economize the approach a vortex field is inserted at the inlet that does not require a precursor simulation or implementation of a cyclic boundary condition at inlet–outlet faces. The controlling parameters are the number of vortices, the size of each vortex, the vorticity (or equivalently velocity field characterizing each vortex), and the lifetime of vortices [16].

Figure 16.2 demonstrates the vortex method, which uses vortices on the inlet boundary to generate velocity fluctuations. The vortices are two dimensional with their vorticity vector parallel to the streamwise direction. The theory is fully

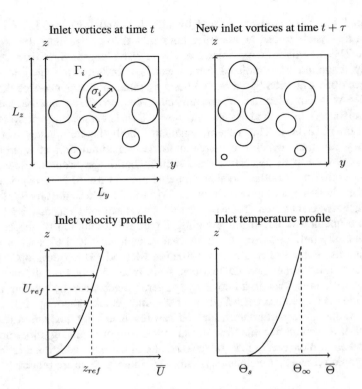

Fig. 16.2 Schematics of vortex generation in the synthetic vortex method

developed in the literature [14–17] and provides the following velocity fluctuation field for a given time step:

$$\mathbf{u}'(\mathbf{x}) = \frac{1}{2\pi} \sum_{i=1}^{N} \Gamma_i \frac{(\mathbf{x}_i - \mathbf{x}) \times \mathbf{s}}{|\mathbf{x}_i - \mathbf{x}|^2} \left(1 - e^{-\frac{|\mathbf{x}_i - \mathbf{x}|^2}{2(\sigma_i(\mathbf{x}_i))^2}} \right) e^{-\frac{|\mathbf{x}_i - \mathbf{x}|^2}{2(\sigma_i(\mathbf{x}_i))^2}}, \qquad (16.20)$$

where $\mathbf{u}'$ [ms^{-1}] is velocity perturbation at the model inlet that is later superimposed on the mean inlet velocity, $\mathbf{x}$ [m] is position vector on the inlet boundary, N [–] is the number of vortices to be inserted at the inlet, i is the index for the current vortex, Γ_i [m^{2}s^{-1}] is the circulation for the current vortex, $\mathbf{x}_i$ [m] is the position vector for the centre of the current vortex, $\mathbf{s}$ [–] is unit vector along the streamwise direction, and $\sigma_i(\mathbf{x}_i)$ [m] is a characteristic length for the radius of current vortex. This formula essentially superimposes velocity fluctuation fields from N [–] vortices to provide an overall perturbation velocity field at the inlet. The specific parameterizations required to develop models for each term in this formula will be provided below.

We assume that the wall-normal direction is $+z$ and that flow is in the $+x$ direction. A power-law profile is assumed for the mean velocity [20, 22] given by:

$$\overline{U}(z) = U_{ref} \left(\frac{z}{z_{ref}} \right)^{\alpha} , \qquad (16.21)$$

where z_{ref} [m] is a reference height, U_{ref} [ms^{-1}] is reference velocity, and α [–] is an exponent parameterized as a function of aerodynamic roughness length. In fact there is a functional relationship between exponent α [–] and the characteristic aerodynamic roughness length of the surface z_0 [m] [22] given as

$$\alpha = \frac{1}{\ln \left(\frac{z_{ref}}{z_0} \right)} . \qquad (16.22)$$

Next a turbulence intensity profile has to be assumed. This is obtained from the relationship:

$$I_u(z) = \frac{1}{\ln \left(\frac{z}{z_0} \right)} , \qquad (16.23)$$

where $I_u(z)$ [–] is limited by a maximum value $I_{u,max}$ [–] given the fact that for atmospheric flows there is a limit to $I_u(z)$ [–] of typically in the order of one [23, 24]. Particularly, with decreasing z_0 [m], the formulation above gives rise to very unrealistically large $I_u(z)$ [–] values near the surface as $z \to 0$ m. This must be avoided by setting the $I_{u,max}$ [–] limit. This allows parameterization of subgrid turbulence kinetic energy k_{sgs} [m^{2}s^{-2}] such that

$$k_{sgs}(z) = 1.5 \left[\overline{U}(z) I_u(z) \right]^2 . \qquad (16.24)$$

To calculate characteristic size for the energy-containing eddies or vortices, we first approximate a characteristic length for the inlet boundary:

$$L = \frac{2 L_z L_y}{L_z + L_y} , \qquad (16.25)$$

where L_z and L_y [m] are inlet height and width. It is reasonable to assume that the size of the largest energy-containing vortices, i.e. σ_{max} [m], scales with L [m] because for atmospheric boundary-layer flow simulations the boundary-layer height δ [m] is in the order of L [m] for economized models. We relate σ_{max} and L [m] using a constant a_σ [–], as a model input variable, with

$$\sigma_{max} = a_\sigma L . \qquad (16.26)$$

For LES, it must be ensured that grid spacing Δ [m] in the coarsest region of mesh, likely on top of the domain, satisfies $\Delta < \sigma_{max}$ [m] [17] since the LES model should be able to resolve the transport, dynamics, and breakdown of the largest eddies in the flow. On the other hand, the size of energy-containing vortices or eddies is a function of height and must decrease with decreasing height. Energy-containing vortex size is parameterized using the mixing length approach of [25] such that

$$\frac{1}{\sigma(z)} = \frac{1}{\sigma_{max}} + \frac{1}{\kappa(z + z_0)} , \tag{16.27}$$

where $\kappa = 0.41$ [–] is the von Kármán constant. This formulation implies that $\sigma(z) \to \kappa z_0$ [m] as $z \to 0$ m and $\sigma(z) \to \sigma_{max}$ [m] as $z \to \infty$ m. It is apparent that $\sigma(z) = \sigma(\mathbf{x})$ [m] is designed to represent the energy-containing eddy size at each height above ground for the synthetic vortex method, and it is incumbent upon the simulation to create the energy cascade, down to the local grid size Δ [m], within a short adaptation distance downstream of the inlet.

A characteristic time for the largest energy-containing vortices or eddies can be approximated using scaling, introduced in Chap. 8. The characteristic velocity U_0 [ms^{-1}] for the largest energy-containing eddies can be defined using the power law and the reference height $U_0 = a z_{ref}^{\alpha}$ [ms^{-1}]. The lengthscale for such eddies can also be found using our definition $\ell_0 = \sigma_{max}$ [m]. These two scales allow calculation of the Reynolds number for the largest energy-containing eddies $Re_{\ell_0} = U_0 \ell_0 / \nu$ [–]. These provide estimates for the Kolmogorov lengthscale $\eta = \ell_0 Re_{\ell_0}^{-3/4}$ [m] (Eq. 8.5), Kolmogorov velocity scale $u_\eta = U_0 Re_{\ell_0}^{-1/4}$ [ms^{-1}] (Eq. 8.6), and dissipation rate $\epsilon = \nu(u_\eta/\eta)^2$ [m^{2}s^{-3}] (Eq. 8.4). This provides the characteristic lifetime for the largest energy-containing eddies in the flow as

$$\tau_0(\ell_0) = \left(\frac{\ell_0^2}{\epsilon} \right)^{1/3} . \tag{16.28}$$

This timescale is not representative for all energy-containing vortices or eddies, but only the largest ones. For ease of implementation, it is possible to define a representative timescale for all energy-containing eddies assuming a constant a_τ [–], as a model input variable, with

$$\tau = a_\tau \tau_0(\ell_0) . \tag{16.29}$$

This timescale can be used to sample a new set of vortices at the inlet after every fixed number of iterations, when this timescale is elapsed.

The circulation can also be parameterized for each vortex knowing the face area S [m^2] of the numerical cell at which a vortex is centred and k_{sgs} [m^{2}s^{-2}] given for a height. The circulation sign is randomized as either positive or negative for each vortex.

$$\Gamma = 4 \left(\frac{\pi \, Sk_{sgs}}{3N(2\ln 3 - 3\ln 2)} \right)^{1/2}. \tag{16.30}$$

For atmospheric boundary-layer flows, an inlet vertical profile for the potential temperature may also be proposed. The potential temperature is the temperature that a parcel of air would attain if adiabatically brought to a standard reference pressure. The power law for potential temperature provides [6]

$$\overline{\Theta}(z) = (\Theta_\infty - \Theta_s) \left(\frac{z}{z_{max}} \right)^\alpha + \Theta_s , \tag{16.31}$$

where Θ_∞ [K] is the potential temperature on top of the model domain z_{max} [m] and Θ_s [K] is the surface potential temperature. Taking account of the variation of potential temperature with height is necessary for non-neutral atmospheric boundary layer.

Figures 16.3, 16.4, 16.5, 16.6, 16.7, and 16.8 show the results of simulations for atmospheric flow in a wind tunnel under weakly and strongly thermally stable conditions [6] using the LES approach of [5, 8], and [6], which use the synthetic inlet turbulence generator and the one-equation turbulence kinetic energy model introduced in this chapter. The concept of thermal stability will be described in detail in Chap. 18. It can be seen that under weakly stable conditions larger flow fluctuations occur and there is a greater degree of mixing in the flow. However, under strongly stable conditions the flow fluctuations are damped, the mixing is reduced, and the flow structure is more stratified.

Problems

16.1 Explain why the Large-Eddy Simulation (LES) is an incomplete turbulence model.

16.2 A computational model domain in Cartesian coordinates in consideration for Large-Eddy Simulation (LES) has uniform mesh spacing with $\Delta x = 4$ m, $\Delta y = 2$ m, and $\Delta z = 1$ m, show that the filter width is calculated as

$$\Delta = 2 \, \text{m} . \tag{16.32}$$

16.3 Some LES models parameterize the residual eddy viscosity $\nu_r(\mathbf{x}, t)$ [m^{2}s^{-1}] independent of the local filter width Δ [m] [26]. Suppose a simple LES model is to be developed with such a formulation of the residual eddy viscosity. For instance, the atmospheric boundary-layer wind is considered. The axis normal to the earth surface is y and wind flows in the x direction parallel to the earth surface. A formulation for residual eddy viscosity is proposed employing a mixing length

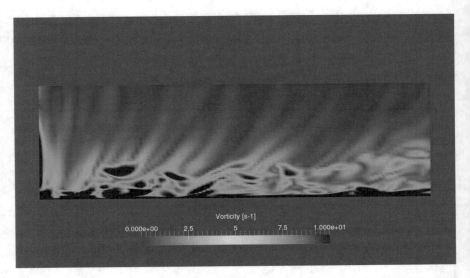

Fig. 16.3 Instantaneous magnitude of vorticity ω [s^{-1}] for wind tunnel flow under weakly thermally stable condition

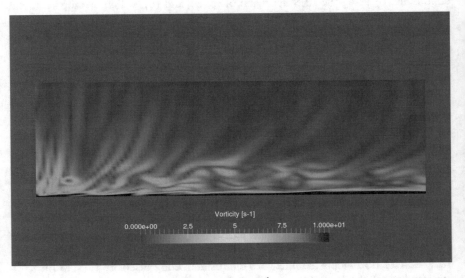

Fig. 16.4 Instantaneous magnitude of vorticity ω [s^{-1}] for wind tunnel flow under strongly thermally stable condition

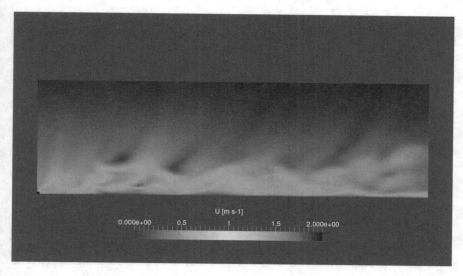

Fig. 16.5 Instantaneous magnitude of velocity U [ms^{-1}] for wind tunnel flow under weakly thermally stable condition

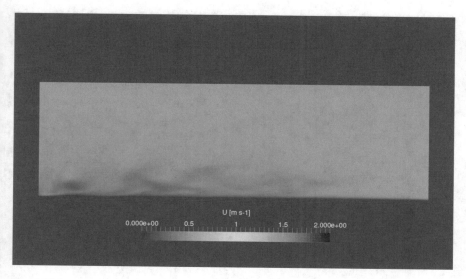

Fig. 16.6 Instantaneous magnitude of velocity U [ms^{-1}] for wind tunnel flow under strongly thermally stable condition

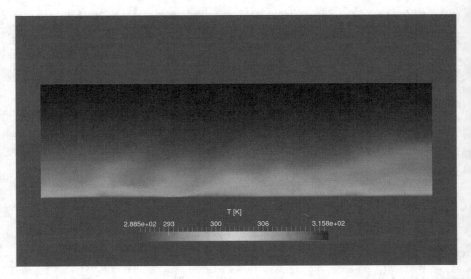

Fig. 16.7 Instantaneous potential temperature Θ [K] for wind tunnel flow under weakly thermally stable condition

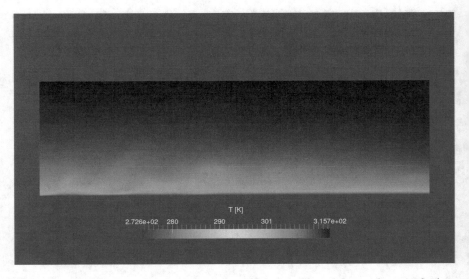

Fig. 16.8 Instantaneous potential temperature Θ [K] for wind tunnel flow under strongly thermally stable condition

relationship, in which the mixing length is only a function of distance away from the earth surface y [m], i.e.

$$v_r = (\ell_m(y))^2 \, \overline{S} \, . \tag{16.33}$$

It is desired to formulate mixing length $\ell_m(y)$ [m] in such a way that it approaches a maximum value ℓ_0 [m] far away from the earth surface, while it approaches κy [m] close to the surface, where $\kappa = 0.41$ [–] is the von Kármán constant, i.e.

$$\begin{cases} y \to 0 & \ell_m(y) \to \kappa y \, , \\ y \to \infty & \ell_m(y) \to \ell_0 \, . \end{cases}$$

Show that the following formulation for $\ell_m(y)$ [m] has this property, where

$$\frac{1}{\ell_m(y)} = \frac{1}{\ell_0} + \frac{1}{\kappa y} \, . \tag{16.34}$$

16.4 Consider the previous problem. The proposed formulation of the mixing length $\ell_m(y)$ [m] was idealistic because near the earth surface, mixing length, and therefore residual eddy viscosity, should approach zero according to this formulation. In other words, the earth surface was assumed absolutely smooth for the first numerical cell such that turbulent flow would vanish in this cell. In practical modelling, however, the first numerical cell adjacent to the earth surface is large enough so that the earth surface is rough in comparison to the characteristic size of the first numerical cell. In this scale, there will be turbulence and hence non-zero mixing length, and therefore non-zero residual eddy viscosity. To remedy this, the mixing length can be parameterized alternatively.

Consider that the roughness of the earth surface can by characterized by lengthscale y_0 [m] such that $y_0 \ll \ell_0$ [m]. It is desired to formulate mixing length $\ell_m(y)$ [m] in such a way that it approaches a maximum value ℓ_0 [m] far away from the earth surface, while it approaches $\kappa(y + y_0)$ [m] close to the surface, where $\kappa = 0.41$ [–] is the von Kármán constant, i.e.

$$\begin{cases} y \to 0 & \ell_m(y) \to \kappa(y + y_0) \, , \\ y \to \infty & \ell_m(y) \to \ell_0 \, . \end{cases}$$

Show that the following formulation for $\ell_m(y)$ [m] has this property, where

$$\frac{1}{\ell_m(y)} = \frac{1}{\ell_0} + \frac{1}{\kappa(y + y_0)} \, . \tag{16.35}$$

16.5 An atmospheric scientist has developed an LES code to simulate the airflow over a flat farm land. Flow parallel to the earth surface is in the direction of positive

x. The direction normal to the earth surface is positive y. The model height is H [m]. The scientist has developed a numerical grid that is uniform in the x and z directions but varies in the y direction. That is, the grid is more refined near the surface but is less refined far away from the surface. The scientist has generated three numerical grids: fine, medium, and coarse. The fine mesh has the smallest cell sizes while the coarse mesh has the largest cell sizes. The local grid size in the domain is given as Δ [m]. It is known that the Kolmogorov scales anywhere in a fluid domain are given by:

$$\eta = \left(\frac{\nu^3}{\epsilon}\right)^{1/4}, \tag{16.36}$$

where ν [m^{2}s^{-1}] is fluid's molecular kinematic viscosity and ϵ [m^{2}s^{-3}] is the local dissipation rate for turbulence kinetic energy. It has been suggested that the maximum dissipation takes place corresponding to a lengthscale of about 24η [m]. Since at least two grid points are needed to resolve a flow feature, a grid spacing of $\Delta = 12\eta$ [m] is required to resolve features of the flow having a scale of 24η [m] [27]. This justifies calculating Δ/η [–] and evaluating where in the domain it is less or greater than 12 [–]. The scientist has produced this plot shown in Fig. 16.9, where non-dimensional height y/H [–] is plotted versus Δ/η [–].

Provide an argument and analyse this plot to determine for each simulation where in the domain the flow features are resolved associated with lengthscale of 24η [m], i.e. the lengthscale associated with maximum dissipation.

Fig. 16.9 LES case: non-dimensional altitude y/H [–] plotted versus the ratio of grid size to Kolmogorov scales Δ/η [–]

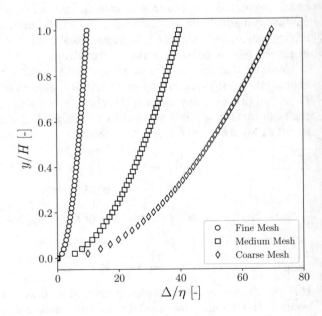

Fig. 16.10 LES case:
non-dimensional altitude
y/H [–] versus ν_r/ν [–]

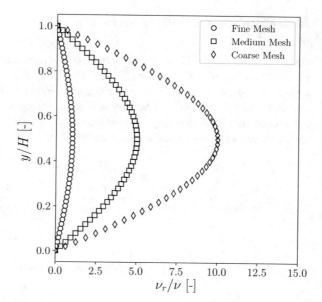

16.6 An atmospheric scientist has developed an LES code to simulate the airflow over an urban area. Flow parallel to the earth surface is in the direction of positive x. The direction normal to the earth surface is positive y. The model height is H [m]. The scientist has generated three numerical grids: fine, medium, and coarse. The fine mesh has the smallest cell sizes while the coarse mesh has the largest cell sizes. It has been suggested that the ratio of the resolved and modelled features of the flow are controlled by the ratio of the residual to molecular viscosities, i.e. ν_r/ν [–] [27]. This justifies calculating this ratio and evaluating where in the domain it is low or high. The scientist has produced this plot shown in Fig. 16.10, where non-dimensional height y/H [–] is plotted versus ν_r/ν [–].

Provide an argument and analyse this plot to determine for each simulation where in the domain the flow features are more resolved than modelled, or where in the domain the flow features are more modelled than resolved. Why does generally a coarser mesh result in more flow features being modelled as opposed to resolved?

16.7 A scientist is developing a Large Eddy Simulation (LES) code for a two-dimensional flow. He is considering a wall flow for a region near the wall given by the following mean velocity profile in units of [ms^{-1}]:

$$\langle U \rangle = \ln y \,, \tag{16.37}$$

in which the x direction is along the wall in the streamwise direction, $\langle U \rangle$ [ms^{-1}] is the mean flow parallel to the wall along the x direction, and y [m] is the direction normal to the wall. Note that the mean velocity normal to the wall is given by $\langle V \rangle = 0$ ms^{-1}. In his model the residual viscosity is given as $\nu_r = 0.2$ m^{2}s^{-1}. Help him calculate the anisotropic residual-stress tensor τ_{xy}^r at $y = 2\,\text{m}$. What is the unit of this quantity?

16.8 A meteorologist is developing a one-equation turbulence kinetic energy model to close the system of equations for a Large-Eddy Simulation (LES). He wishes to add the temperature and passive scalar equations to the continuity, momentum, and Sub-Grid Scale (SGS) turbulence kinetic energy equations. For this model, the momentum, temperature, and passive scalar equations for the filtered velocity, temperature, and passive scalar are given by [28]:

$$\underbrace{\frac{\overline{D}\,\overline{U}_i}{\overline{Dt}}}_{\text{Mean Substantial Derivative}} = \underbrace{\nu\frac{\partial^2\overline{U}_i}{\partial x_j\partial x_j}}_{\text{Surface Forces}} + \underbrace{\frac{\partial}{\partial x_j}\left(2\nu_r\overline{S}_{ij}\right)}_{\text{Anisotropic Residual} - \text{Stress Forces}} \qquad (16.38)$$

$$- \underbrace{\frac{1}{\rho}\frac{\partial\overline{p}^m}{\partial x_i}}_{\text{Modified Pressure Forces}} - \underbrace{g\delta_{i3}}_{\text{Buoyancy Body Force}} ,$$

$$\underbrace{\frac{\overline{D}\,\overline{T}}{\overline{Dt}}}_{\text{Mean Substantial Derivative}} = \underbrace{\alpha_\theta\frac{\partial^2\overline{T}}{\partial x_j\partial x_j}}_{\text{Molecular Heat Diffusion}} + \underbrace{\frac{\partial}{\partial x_j}\left(\alpha_{\theta r}\frac{\partial\overline{T}}{\partial x_j}\right)}_{\text{Turbulent Heat Diffusion}} , \qquad (16.39)$$

$$\underbrace{\frac{\overline{D}\,\overline{S}}{\overline{Dt}}}_{\text{Mean Substantial Derivative}} = \underbrace{\alpha_S\frac{\partial^2\overline{S}}{\partial x_j\partial x_j}}_{\text{Molecular Passive Scalar Diffusion}} + \underbrace{\frac{\partial}{\partial x_j}\left(\alpha_{Sr}\frac{\partial\overline{S}}{\partial x_j}\right)}_{\text{Turbulent Passive Scalar Diffusion}} .$$
$$(16.40)$$

Note that the buoyancy body force due to gravitational acceleration is not absorbed in the modified pressure force term in this formulation. The term δ_{i3} [–] ensures that the gravitational acceleration is only considered in the vertical direction, i.e. z direction or the third component of the Cartesian coordinate system. ν and ν_r [m^2s^{-1}] are molecular and residual viscosities. α_θ and $\alpha_{\theta r}$ [m^2s^{-1}] are molecular and residual thermal diffusivities. Likewise, α_S and α_{Sr} [m^2s^{-1}] are molecular and residual passive scalar diffusivities. The turbulence kinetic energy model is given by:

$$\underbrace{\frac{\overline{D}k_{sgs}}{\overline{Dt}}}_{\text{Mean Substantial Derivative}} = \underbrace{\nabla\cdot\left(\frac{\nu_r}{\sigma_{k_{sgs}}}\nabla k_{sgs}\right)}_{\text{Energy Flux Divergence}} + \underbrace{\mathcal{P}}_{\text{Shear Production}}$$

$$+ \underbrace{\mathcal{B}}_{\text{Buoyant Production}} - \underbrace{\epsilon}_{\text{Dissipation}} , \qquad (16.41)$$

where $\mathcal{P}$ and $\mathcal{B}$ [m^{2}s^{-3}] account for shear and buoyant production of turbulence kinetic energy, respectively. Here, the SGS Turbulence Kinetic Energy (TKE) k_{sgs} [m^{2}s^{-2}] does not represent fluctuations at all scales but only at smallest scales below the filter width Δ [m]. $\sigma_{k_{sgs}}$ [–] is the turbulent Prandtl number for SGS TKE. The production terms are closed as

$$\mathcal{P} = 2\nu_r \left(\overline{S}_{ij}\right)^2 , \tag{16.42}$$

$$\mathcal{B} = -\frac{g\alpha_{\theta r}}{T_0} \frac{\partial \overline{T}}{\partial z} , \tag{16.43}$$

where buoyant production is only related to vertical gradient of the resolved temperature. The residual viscosity ν_r is closed using the following relationship employing the SGS TKE k_{sgs} [m^{2}s^{-2}] and more constants:

$$\nu_r = C_k k_{sgs}^{1/2} l . \tag{16.44}$$

Here $C_k = 0.094$ [–] is a constant, and $l = C_\Delta (\Delta x \Delta y \Delta z)^{1/3}$ [m] is the subgrid mixing length. Constant C_Δ [–] is usually in the order of one but can be optimized to control the SGS dissipation rate of turbulence kinetic energy. Finally the dissipation rate of turbulence kinetic energy ϵ [m^{2}s^{-3}] is closed using one more constant

$$\epsilon = C_\epsilon \frac{k_{sgs}^{3/2}}{l} , \tag{16.45}$$

with $C_\epsilon = 1.048$ [–]. As noted above the turbulence model is not closed yet, because further parameterizations are necessary for α_θ, $\alpha_{\theta r}$, α_S, and α_{Sr} [m^{2}s^{-1}]. Provide simple parameterizations for these terms as functions of either ν or ν_r [m^{2}s^{-1}].

References

1. Smagorinsky J (1963) General circulation experiments with the primitive experiments: I. The basic equations. Mon Weather Rev 91:99–164
2. Deardorff J W (1974) Three-dimensional numerical study of the height and mean structure of a heated planetary boundary layer. Bound-Lay Meteorol 7:81–106
3. Talbot C, Bou-Zeid E, Smith J (2012) Nested mesoscale large-eddy simulations with WRF: performance in real test cases. J Hydrometeorol 13:1421–1441
4. Nahian M R, Nazem A, Nambiar M K et al (2020) Complex meteorology over a complex mining facility: Assessment of topography, land use, and grid spacing modifications in WRF. J Appl Meteorol Clim 59:769–789
5. Aliabadi A A, Veriotes N, Pedro G (2018) A Very Large-Eddy Simulation (VLES) model for the investigation of the neutral atmospheric boundary layer. J Wind Eng Ind Aerodyn 183:152–171

6. Ahmadi-Baloutaki M, Aliabadi A A (2021) A very large-eddy simulation model using a reductionist inlet turbulence generator and wall modeling for stable atmospheric boundary layers. Fluid Dyn 56:413–432

7. Li X-X, Britter R E, Koh T Y, et al (2010) Large-eddy simulation of flow and pollutant transport in urban street canyons with ground heating. Bound-Lay Meteorol 137:187–204

8. Aliabadi A A, Krayenhoff E S, Nazarian N et al (2017) Effects of roof-edge roughness on air temperature and pollutant concentration in urban canyons. Bound-Lay Meteorol 164:249–279

9. Tabor G R, Baba-Ahmadi M H (2010) Inlet conditions for large eddy simulation: A review. Comput Fluids 39:553–567

10. Castro H G, Paz R R (2013) A time and space correlated turbulence synthesis method for large eddy simulations. J Comput Phys 235:742–763

11. Lund T S, Wu X, Squires K D (1998) Generation of turbulent inflow data for spatially-developing boundary layer simulations. J Comput Phys 140:233–258

12. Spalart P R (1988) Direct simulation of a turbulent boundary layer up to $R_\theta = 1410$. J Fluid Mech 187:61–98

13. Cao S (2014) Advanced physical and numerical modeling of atmospheric boundary layer. Journal of Civil Engineering Research 4:14–19

14. Sergent M E (2002) Vers une méthodologie de couplage entre la simulation des grandes echelles et les modéles statistique. École Centrale De Lyon, Écully

15. Benhamadouche S, Jarrin N, Addad Y et al (2006) Synthetic turbulent inflow conditions based on a vortex method for large-eddy simulation. Prog Comput Fluid Dy 6:50–57

16. Mathey F, Cokljat D, Bertoglio J P et al (2006) Assessment of the vortex method for large eddy simulation inlet conditions. Prog Comput Fluid Dy 6:58–67

17. Xie B (2016) Improved vortex method for LES inflow generation and applications to channel and flat-plate flows. École Centrale De Lyon, Écully

18. Aboshosha H, Bitsuamlak G, El Damatty A (2015) LES of ABL flow in the built-environment using roughness modeled by fractal surfaces. Sustain Cities Soc 19:40–60

19. Castro H G, Paz R R, Mroginski J L et al (2017) Evaluation of the proper coherence representation in random flow generation based methods. J Wind Eng Ind Aerodyn 168:211–227

20. Ricci M, Patruno L, de Miranda S (2017) Wind loads and structural response: Benchmarking LES on a low-rise building. Eng Struct 144:26–42

21. Buckingham S, Koloszar L, Bartosiewicz Y et al (2017) Optimized temperature perturbation method to generate turbulent inflow conditions for LES/DNS simulations. Comput Fluids 154:44–59

22. Thomas T G, Williams J J R (1999) Generating a wind environment for large eddy simulation of bluff body flows. J Wind Eng Ind Aerodyn 82:189–208

23. Stull R B (1988) An introduction to boundary layer meteorology. Kluwer Academic Publishers, Dordrecht

24. Nozawa K, Tamura T (2002) Large eddy simulation of the flow around a low-rise building immersed in a rough-wall turbulent boundary layer. J Wind Eng Ind Aerodyn 90:1151–1162

25. Mellor G L, Yamada T (1974) A hierarchy of turbulence closure models for planetary boundary layers. J Atmos Sci 31:1791–1806

26. Mason P J, Callen N S (1986) On the magnitude of the subgrid-scale eddy coefficient in large-eddy simulations of turbulent channel flow. J Fluid Mech 162:439–462

27. Fröhlich J, Mellen C P, Rodi W et al (2005) Highly resolved large-eddy simulation of separated flow in a channel with streamwise periodic constrictions. J Fluid Mech 526:19–66

28. Kia S, Flesch T K, Freeman B S et al (2021) Atmospheric transport over open-pit mines: The effects of thermal stability and mine depth. J Wind Eng Ind Aerodyn 214:104677

Chapter 17
Direct Numerical Simulation

Abstract This chapter briefly introduces the Direct Numerical Simulation (DNS) technique, which directly simulates all scales of turbulent motion. DNS does not require any turbulence modelling. In this chapter, simple scaling analyses are performed to estimate the computational expense associated with DNS. This involves calculating the number of spatial grid elements and the number of time iterations required for a given simulation.

17.1 Overview

For the first time [1] showed that it is possible to perform computer simulations of a fully developed turbulent flow without the need to model or parameterize turbulence in a closure-like approach. In the so-called Direct Numerical Simulation (DNS) every eddy from the largest to the smallest is computed. In this approach the Navier–Stokes equations are integrated forward given some initial and boundary conditions. Using this approach the entire velocity field is available at all spatial and temporal scales of turbulence [2].

Although attractive, there are serious computational limits in using DNS given today's computer technology. It was discussed earlier that the Kolmogorov length-scale in turbulent flow scales with $\eta \sim Re^{-3/4}\ell_0$ [m] (Eq. 8.5), where ℓ_0 [m] is the size of energy-containing eddies and Re [−] is the Reynolds number of the flow. If a DNS shall resolve all eddies, then the numerical grid must be as spatially refined as the Kolmogorov lengthscale. In other words, given the Cartesian coordinate, the grid resolution should satisfy

$$\Delta x \sim \Delta y \sim \Delta z \sim Re^{-3/4}\ell_0 . \tag{17.1}$$

This requirement can quickly become inhibiting with increasing Reynolds number. The number of grid points required at any instant for a three-dimensional simulation is therefore

$$N_x \sim \left(\frac{L_{\text{box}}}{\Delta x}\right)^3 \sim \left(\frac{L_{\text{box}}}{\ell_0}\right)^3 Re^{9/4} , \tag{17.2}$$

where L_{box} [m] is a typical dimension of the computational domain. We can immediately spot the problem, that a large Reynolds number requires a large number of grid points for a practical DNS. Furthermore, we can assess the number of time steps required for a DNS given the Kolmogorov lengthscale. The maximum permissible time step in a simulation is in the order of $\Delta t \sim \Delta x / \langle U \rangle \sim \eta / \langle U \rangle$ [s] since, in order to maintain numerical stability and accuracy, we cannot allow a fluid parcel travel more than on grid spacing per time step. If T [s] is the total duration of the simulation, then the minimum number of time steps required for a DNS is

$$N_t \sim \frac{T}{\Delta t} \sim \frac{T}{\eta / \langle U \rangle} \sim \frac{T}{\ell_0 / \langle U \rangle} Re^{3/4} . \tag{17.3}$$

The number of computer operations required for DNS is thus obtained by multiplying N_x [−] and N_t [−], so the computational cost scales with

$$\text{Computational Cost} \sim N_x N_t \sim \left(\frac{T}{\ell_0 / \langle U \rangle}\right)\left(\frac{L_{\text{box}}}{\ell_0}\right)^3 Re^3 . \tag{17.4}$$

Given today's computational power, size of the domain, and the Reynolds number, a DNS for a practical problem may take anywhere between few hours to centuries to run. At the moment DNS investigators have spent most of the effort on simple possible geometries such as a periodic cube or box turbulence [3]. Someday with the advent of quantum computing, it may be possible to apply DNS to more and more practical problems with higher domains and larger Reynolds numbers.

Problems

17.1 Assuming that $L_{\text{box}} = 1$ m and that $Re = 10,000$ [−], estimate the number of spatial grid points for a three-dimensional DNS. How did you estimate ℓ_0 [m]?

17.2 Many practical DNS models do not use Kolmogorov length and timescales to determine the grid resolution or time step. Instead they choose larger length and timescales that correspond to the finest scales of the inertial subrange [4]. This way, turbulent fluctuations finer than these scales are ignored and models can perform simulations more computationally efficiently. In the inertial subrange, the velocity and timescales for an eddy of size ℓ [m] are given by:

$$u(\ell) = (\epsilon \ell)^{1/3} , \tag{17.5}$$

$$\tau(\ell) = (\ell^2 / \epsilon)^{1/3} , \tag{17.6}$$

where ϵ [m^{2}s^{-3}] is the dissipation rate of the turbulence kinetic energy. If one rearranges the expressions for the velocity and timescales, one can obtain the eddy length and timescales as functions of velocity scale and dissipation rate. The eddy length and timescales can be used to estimate the computational cost of a DNS model assuming that the grid resolution is in the order of ℓ [m] and simulation time step is in the order of $\tau(\ell)$ [s]. Assuming that L_{box} [m] is a typical dimension of the computational domain and that T [s] is the total duration of the simulation, show that the computational cost scales with

$$\text{Computational Cost} \sim N_x N_t \sim L_{\text{box}}^3 T \frac{\epsilon^4}{u(\ell)^{11}} . \tag{17.7}$$

References

1. Orszag S A, Patterson G S (1972) Numerical simulation of three-dimensional homogeneous isotropic turbulence. Phys Rev Lett 28:76–79
2. Pope S B (2000) Turbulent flows. Cambridge University Press, Cambridge
3. Davidson P A (2005) Turbulence: an introduction for scientists and engineers. Oxford University Press, Oxford
4. Coceal O, Thomas T G, Castro I P et al (2006) Mean flow and turbulence statistics over groups of urban-like cubical obstacles. Bound-Lay Meteorol 121:491–519

Chapter 18
Wall Models

Abstract This chapter introduces the concept of wall modelling, which helps economize numerical simulations of fluid flows that involve fluid interaction with one or multiple surfaces. Wall modelling reduces the required number of spatial grid elements in the domain near a surface, thus significantly reducing the computational cost. The point-wise standard wall function is introduced, which develops expressions for flow properties in the log-law region near a wall at a specific point. The integrated Werner–Wengle wall function is introduced, which develops expressions for flow properties in the linear and power-law regions near a wall integrated over a region of space from the surface to a location in the power-law regime. The van Driest near-wall treatment is described, which is typically employed in LES. Finally, wall models for the atmospheric boundary layer are developed, which are described using the Monin–Obukhov Similarity Theory (MOST), considering the atmospheric thermal stability condition, surface roughness lengthscales, and displacement height.

18.1 Preliminaries

The near-wall region in a turbulent flow adds complexity and computational expense to the task of performing calculations for accurately predicting turbulent flows. Given the very steep profiles for most solution variables near the wall, such as k [m^2s^{-2}], ϵ [m^2s^{-3}], $\langle U \rangle$ [ms^{-1}], etc., resolving the near-wall regions in a turbulent simulation is very costly. As a result, alternative approaches must be found to model or approximate the solution behaviour near the walls using some form of algebraic models, without having to resolve the solution profiles in detail near the walls.

The *wall function* approach, first introduced by [1], applies boundary conditions at some distance away from the wall, so that turbulence model equations do not have to be solved close to the wall, i.e. between the wall and the location at which the boundary conditions are applied. Generally, if the first computational cell adjacent to a wall lies entirely inside the viscous sublayer, then a wall function is not required. However, if the first computational cell covers parts of the buffer layer and beyond, then a wall function is necessary for accurate simulation of turbulence. Definitions

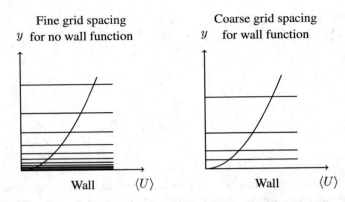

Fig. 18.1 Velocity profile and spatial grid spacing near a wall for a fluid simulation without a wall function (left) and a fluid simulation with a wall function (right); the grid spacing for a simulation without a wall function shall be sufficiently fine to enable calculation of flow properties near a wall with very steep profiles; the grid spacing for a simulation with a wall function can be coarser since wall boundary conditions are applied at a distance away from the wall

for near-wall flow layers can be found in Chap. 5. Figure 18.1 shows the grid spacing configuration near a wall for a fluid simulation that does not employ a wall function and a simulation that utilizes a wall function. It is evident that using a wall function reduces the required resolution of the grid spacing near a wall.

18.2   Point-Wise Standard Wall Function

Figure 18.2 shows the placement of the first computational cell with respect to wall regions according to the law of the wall introduced in Chap. 5. The entire height of the first computational cell is $\Delta y = 2y_p$ [m], where y_p [m] is the distance from wall to the centre of the computational cell. The *standard wall function* assumes that point y_p [m] is located in the log-law region. The wall function boundary conditions are applied at this location, i.e. $y = y_p$ [m]. The subscript p indicates quantities evaluated at y_p [m], e.g. $\langle U \rangle_p$ [ms^{-1}], k_p [m^{2}s^{-2}], and ϵ_p [m^{2}s^{-3}]. For a high-Reynolds-number zero-pressure-gradient boundary layer, the log-law Eq. 5.20 is given as

$$u^+ \equiv \frac{\langle U \rangle}{u_\tau} = \frac{1}{\kappa} \ln y^+ + B \ . \tag{18.1}$$

It must be remembered that $u_\tau \equiv \sqrt{\tau_w/\rho}$ [ms^{-1}] is defined as the *friction velocity*, where τ_w [kgm^{-1}s^{-2}] is the shear stress at the wall. The balance of the production rate and dissipation rate of turbulence kinetic energy near the wall yields

$$\epsilon = \frac{u_\tau^3}{\kappa y} \ . \tag{18.2}$$

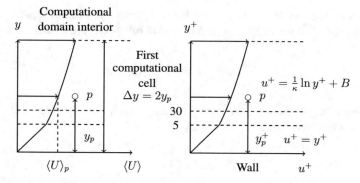

Fig. 18.2 Placement of the first computational cell with respect to wall regions according to the law of the wall

In addition, using the $k - \epsilon$ model introduced in Chap. 15, and Eq. 15.17 for the turbulent viscosity for simple shear flows, it is possible to relate shear stress near the wall to the turbulence kinetic energy by:

$$- \langle uv \rangle = u_\tau^2 = C_\mu^{1/2} k \, . \tag{18.3}$$

The *standard wall function* uses these relations to provide robust boundary conditions under all circumstances at location y_p [m]. First a *nominal friction velocity* is defined using the value of turbulence kinetic energy at distance y_p [m], i.e. k_p [m²s⁻²]:

$$u_\tau^* \equiv C_\mu^{1/4} k_p^{1/2} \, . \tag{18.4}$$

In turbulent simulations, an exact value of $y_p^+ \equiv u_\tau y / v \, [-]$ is not known since an accurate estimate of u_τ [ms⁻¹] is not available. However, once the nominal friction velocity is calculated, it is possible to estimate the corresponding $y_p^+ \, [-]$ by the following relationship:

$$y_p^* \equiv \frac{y_p u_\tau^*}{v} \, . \tag{18.5}$$

The nominal mean velocity is then obtained from the log-law relationship, which can approximate the true mean velocity at position y_p [m], which lies in the log-law region

$$\langle U \rangle_p^* = u_\tau^* \left(\frac{1}{\kappa} \ln y_p^* + B \right) \, . \tag{18.6}$$

The boundary condition at y_p [m] for the mean momentum equation is not applied by specifying a $\langle U \rangle_p$ [ms^{-1}] but instead by specifying a shear stress as

$$- \langle uv \rangle_p = u_\tau^{*2} \frac{\langle U \rangle_p}{\langle U \rangle_p^*} . \qquad (18.7)$$

The boundary condition for ϵ [m^{2}s^{-3}] can be conveniently defined having the nominal friction velocity as

$$\epsilon_p = \frac{u_\tau^{*3}}{\kappa y_p} , \qquad (18.8)$$

while zero-normal gradient conditions are applied to k [m^{2}s^{-2}] and to the normal stresses. In finite volume simulations of turbulent flow, the location of y_p [m] is taken to be the first grid node away from the wall. Wall functions in general introduce y_p [m] as an artificial parameter. For boundary layer flows for which the log-law relations are accurate, the overall solution is insensitive to the choice of y_p [m], as long as it is within the log-law region. However, in other flows it is found that the overall solution is sensitive to this choice. As a result, it may not be possible to obtain numerically accurate and grid-independent solutions, since refining grids usually means reducing y_p [m].

18.3   Integrated Werner–Wengle Wall Function

As was seen in the standard wall function, one needs to solve and resolve k [m^{2}s^{-2}] near the wall, at least at point y_p [m], to be able to estimate the friction velocity by the nominal friction velocity. For the wall function to provide the shear stress boundary condition at point y_p [m] in this way requires an iterative approach that could impose computational cost. Alternatively, a wall function was proposed by [2] that eliminated the need for a solution of k [m^{2}s^{-2}] because it enabled a closed form solution for the shear stress near the wall given other known parameters. In addition, this wall function integrated the entire profile of u^+ over the first computational cell in the direction normal to the wall to arrive at a better estimate of u^+ [−] compared to point-wise models. The *Werner–Wengle wall function* provides the following relationship between u^+ [−] and y^+ [−]:

$$\begin{cases} u^+ = y^+ & \text{if } y^+ \leq 11.81 , \\ u^+ = A(y^+)^B & \text{if } y^+ > 11.81 , \end{cases}$$

where $A = 8.3$ [−] and $B = 1/7$ [−]. The two near-wall regions intersect at the value of $y^+ = 11.81$ [−] known as the intersecting y^+ [−] or y_i^+ [−], which can be given in terms of A [−] and B [−] as

$$y_i^+ = A^{\frac{1}{1-B}} . \tag{18.9}$$

It is possible to find an average value for u^+ [−] over the entire computational cell in the wall normal direction, i.e. from $y^+ = 0$ [−] to $y^+ = \Delta y^+ = 2y_p^+$ [−], by performing the following integral:

$$
\begin{aligned}
u_{avg}^+ &= \frac{1}{2y_p^+} \int_0^{2y_p^+} u^+(y^+) dy^+ \\
&= \frac{1}{2y_p^+} \left(\int_0^{y_i^+} y^+ dy^+ + \int_{y_i^+}^{2y_p^+} A(y^+)^B dy^+ \right) ,
\end{aligned}
\tag{18.10}
$$

where the integral is split into two integrals appropriate for each near-wall region. This definite integral can be evaluated such that

$$2u_{avg}^+ y_p^+ = \frac{y_i^{+2}}{2} + \frac{A}{1+B} \left((2y_p^+)^{1+B} - (y_i^+)^{1+B} \right) , \tag{18.11}$$

which by substitution of u_{avg}^+ [−], y_p^+ [−], and y_i^+ [−] can be re-expressed as

$$2\frac{\langle U \rangle_{avg}}{u_\tau} \frac{y_p u_\tau}{\nu} = \frac{1}{2} A^{\frac{2}{1-B}} + \frac{A}{1+B} \left(\left(\frac{2y_p u_\tau}{\nu} \right)^{1+B} - A^{\frac{1+B}{1-B}} \right) . \tag{18.12}$$

Note that $\langle U \rangle_{avg}$ [ms^{-1}] represents the control volume-averaged mean velocity, not to be confused with $\langle U \rangle_p$ [ms^{-1}], which represented the mean velocity at point p in the standard wall function [3]. This formula can be rearranged to give the square of friction velocity, or alternatively wall shear stress, i.e. $-\langle uv \rangle = u_\tau^2$ [m^{2}s^{-2}], as an explicit function of other known variables such as $\langle U \rangle_{avg}$ [ms^{-1}], y_p [m], A [−], B [−], and ν [m^{2}s^{-1}], without the need for knowledge of k [m^{2}s^{-2}], such that

$$-\langle uv \rangle_{avg} = \left[\frac{1+B}{A} \left(\frac{\nu}{2y_p} \right)^B \langle U \rangle_{avg} + \left(\frac{\nu}{2y_p} \right)^{1+B} \left(A^{\frac{1+B}{1-B}} - \frac{1+B}{2A} A^{\frac{2}{1-B}} \right) \right]^{\frac{2}{1+B}} . \tag{18.13}$$

Note that in control volume schemes for computational fluid dynamics analysis, most solution variables are in fact averaged throughout each control volume. The Werner–Wengle wall function is consistent with control volume schemes such that the boundary condition for the wall shear stress in the mean momentum equation for the first computational cell adjacent to the wall can be conveniently applied.

18.4   van Driest Near-Wall Treatment

In LES based on the Smagorinsky closure scheme introduced in Chap. 16 [4], the *Smagorinsky lengthscale* was given as $\ell_S = C_S \Delta$ [m] in Eq. 16.14, where C_S [−] is a constant and Δ [m] is filter length, typically chosen as the geometric average grid size, i.e. $\Delta = (\Delta x \Delta y \Delta z)^{1/3}$ [m] (Eq. 16.15). However, near walls the actual lengthscale ℓ_S [m] can reduce more significantly than calculated using the Smagorinsky lengthscale formula. If this reduction is not taken into account, then excessive dissipation of turbulence kinetic energy near the walls may be predicted. To prevent this, [5] suggested that ℓ_S [m] can be damped near walls according to the *van Driest function*:

$$\ell_S = C_S \Delta \left[1 - \exp\left(\frac{-y^+}{A^+} \right) \right], \tag{18.14}$$

with a typical value for $A^+ = 25$ [−]. Note that for LES the van Driest function can be used in combination with a wall function such as the Werner–Wengle model to result in more accurate simulations without having to resolve turbulence near the walls.

18.5   Wall Models for the Atmospheric Boundary Layer

Atmospheric flows near the earth surface are another example of wall flows. To develop wall models for atmospheric boundary layer flows, more considerations are necessary, involving incorporation of the roughness lengthscale for the surface, fluxes of momentum, heat, and humidity, as well as the thermal stability into the wall model.

Figure 18.3 shows the structure of the entire atmospheric boundary layer demonstrated by vertical profiles of virtual potential temperature and wind speed under different thermal stability and atmospheric conditions. Virtual potential temperature is the theoretical potential temperature (Eq. 15.31) of dry air that would have the same density as moist air. For virtual potential temperature, the profile may exhibit a combination of features such as a constant-value layer, stratified layer, a residual layer, and an inversion point, depending on the thermal stability and atmospheric conditions. The wind speed profile may also monotonically increase, approximated by the power or log laws, or it may exhibit localized jets, again given thermal stability and the atmospheric conditions [6, 7].

Formulation of wall models for the atmospheric boundary layers is achieved using the Monin–Obukhov Similarity Theory (MOST) [9, 10]. MOST is successful in describing the physics of the surface layer, also known as the constant flux layer. This layer of air is in the order of tens of metres thick adjacent to the ground

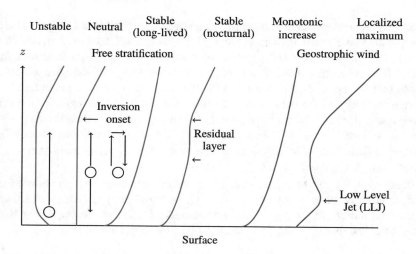

Fig. 18.3 Atmospheric planetary boundary layer schematic showing thermally stable, neutral, and unstable conditions (wind speed (blue lines), virtual potential temperature (red lines), LLJ: Low Level Jet). In the unstable condition, an adiabatic parcel of air (shown as a black circle), if displaced from an original position near the surface, will accelerate upward until reaching the first inversion onset; the same parcel under neutral conditions will not accelerate; under stable cases, the parcel will experience a resisting force due to stratification pushing it back toward the original position [8]

where mechanical (shear) generation of turbulence exceeds buoyant generation or consumption.

In atmospheric boundary layer flows, it is typically assumed that the coordinate direction normal to the earth surface is z (not y). We next introduce the concept of roughness lengthscale for the earth surface. There are three types of roughness lengthscale: aerodynamic roughness z_0 [m], roughness for temperature z_T [m], and roughness for specific humidity z_Q [m]. The aerodynamic roughness lengthscale z_0 [m] is typically a fraction of the average obstacle height (e.g. desert sand, grass, trees, buildings, etc.), typically in the order of 10% of the average obstacle height, on the earth surface. Reference [11] defines different values for z_0 [m] in a comprehensive study of the rough-wall boundary layers. For roughness length for temperature and specific humidity, it is typically assumed $z_\Theta = z_Q = 0.1 z_0$ [m] [12–15].

We next introduce the concept of atmospheric thermal stability. Thermal stability is usually expressed using the non-dimensional stability parameter $\frac{z}{L}$ [−], where z [m] is height above ground and L [m] is Obukhov length, which expresses the relative roles of shear and buoyancy in the production/consumption of turbulence kinetic energy. L [m] is defined as

$$L = -\frac{\Theta_0 u_\tau^3}{\kappa g \langle w\theta \rangle} , \qquad (18.15)$$

where Θ_0 [K] is the reference potential temperature, u_τ [ms^{-1}] is friction velocity, κ [−] is the von Kármán constant, g [ms^{-2}] is gravitational acceleration, and $\langle w\theta \rangle$ [Kms^{-1}] is the vertical sensible turbulent kinematic heat flux. Positive heat flux means that the earth surface is warming the atmosphere, and negative heat flux means that the earth surface is cooling the atmosphere. It must be noted that both friction velocity and the heat flux are measured in the turbulent region of the flow at elevations from 2 to 10 m above ground. Thermally stable conditions occur when $\langle w\theta \rangle < 0$ Kms^{-1} or $\frac{z}{L} > 0$ [−], and thermally unstable conditions occur when $\langle w\theta \rangle > 0$ Kms^{-1} or $\frac{z}{L} < 0$ [−]. Near neutral conditions occur when $\langle w\theta \rangle \to 0$ Kms^{-1}, which implies $\frac{z}{L} \to 0$ [−].

Finally MOST can be introduced, which establishes relationships between the vertical gradient of wind speed $\langle U \rangle$ [ms^{-1}], potential temperature $\langle \Theta \rangle$ [K], and specific humidity $\langle Q \rangle$ [kgkg^{-1}] and corresponding universal functions for these three variables, $\Phi_M\left(\frac{z}{L}\right)$, $\Phi_H\left(\frac{z}{L}\right)$, and $\Phi_Q\left(\frac{z}{L}\right)$ [−], respectively, given by:

$$\frac{d\langle U \rangle}{dz} = \frac{u_\tau}{\kappa z}\Phi_M\left(\frac{z}{L}\right) , \tag{18.16}$$

$$\frac{d\langle \Theta \rangle}{dz} = -\frac{Q_{sen}}{\rho c_p \kappa u_\tau z}\Phi_H\left(\frac{z}{L}\right) , \tag{18.17}$$

$$\frac{d\langle Q \rangle}{dz} = -\frac{Q_{lat}}{\rho L_v \kappa u_\tau z}\Phi_Q\left(\frac{z}{L}\right) , \tag{18.18}$$

where $Q_{sen} = \rho c_p \langle w\theta \rangle$ [Wm^{-2}] is the vertical dynamic sensible heat flux per unit area, and $Q_{lat} = \dot{m}L_v$ [Wm^{-2}] is the vertical latent heat flux per unit area, in which $\dot{m}$ [kgs^{-1}m^{-2}] is water evaporation rate per unit area and L_v [Jkg^{-1}] is the latent heat of evaporation. The universal functions have been discovered empirically [16, 17]. In addition, given the similarity of heat and mass transfer, it may be assumed that $\Phi_H\left(\frac{z}{L}\right) = \Phi_Q\left(\frac{z}{L}\right)$ [−] [18], such that [19]

$$\Phi_M\left(\frac{z}{L}\right) = \begin{cases} 1 + 5\frac{z}{L}, & \frac{z}{L} > 0 (\text{Stable}) , \\ 1, & \frac{z}{L} = 0 (\text{Neutral}) , \\ \left(1 - \frac{16z}{L}\right)^{-1/4}, & \frac{z}{L} < 0 (\text{Unstable}) , \end{cases} \tag{18.19}$$

$$\Phi_H\left(\frac{z}{L}\right) = \Phi_Q\left(\frac{z}{L}\right) = \begin{cases} 1 + 5\frac{z}{L}, & \frac{z}{L} > 0 (\text{Stable}) , \\ 1, & \frac{z}{L} = 0 (\text{Neutral}) , \\ \left(1 - \frac{16z}{L}\right)^{-1/2}, & \frac{z}{L} < 0 (\text{Unstable}) . \end{cases} \tag{18.20}$$

To obtain the vertical profiles of atmospheric variables, the differential equations above should be integrated. This integration is usually made from the roughness length, i.e. z_0 [m] for momentum, z_T [m] for temperature, and z_Q [m] for specific humidity, to altitude z [m]. The integration of the relationships above is straightforward for the thermally stable and neutral cases. However, for the thermally unstable case more mathematical rigor is required [20]. Sometimes the altitude, up to which the integration is taken, is corrected using a zero displacement height d [m] based on the premise that tall obstacle heights shift the profile of atmospheric variables up by amount d [m] that is usually a percentage of the average obstacle height. Comprehensive relationships for d [m] are provided by [11] and [21]. If this is considered, the integration should be performed from z_0 [m] for momentum, z_T [m] for temperature, and z_Q [m] for specific humidity, to altitude $z - d$ [m].

18.6 Wall Function Summary

Wall functions are widely used to economize turbulent simulations. However, they can be much more complex than introduced here. For instance, other wall laws may be assumed, such as multi-layer laws, considering more than two layers discussed here [22]. Another complexity that arises is wall modelling for very rough walls with possibly vertical and horizontal heterogeneous roughness structures. For such cases the law of the wall may change and require application-specific wall models [11, 23–27].

Problems

18.1 For the Werner–Wengle wall function [2] show that

$$y_i^+ = A^{\frac{1}{1-B}} .$$
(18.21)

18.2 Suppose that the Werner–Wengle wall function [2] is to be used in the point-wise form, i.e. the profile of u^+ [−] is not supposed to be integrated, but instead the value of the shear stress $\langle uv \rangle_p$ [m²s⁻²] at y_p [m] is desired as a function of $\langle U \rangle_p$ [ms⁻¹], v [m²s⁻¹], y_p [m], A [−], and B [−]. If point y_p [m] is located in the power-law region, show that

$$- \langle uv \rangle_p = u_\tau^2 = \left(\frac{v^B \langle U \rangle_p}{A y_p^B} \right)^{\frac{2}{1+B}} .$$
(18.22)

18.3 In Computational Fluid Dynamics (CFD) models that use wall functions, it is very critical for the wall law to be valid in the majority of the spatial extent of the first computational cell adjacent to the wall. The coarser the first computational cell adjacent to the wall, the higher the likelihood that the wall law may not satisfy this requirement. If an incompetent CFD modeller uses a very coarse wall-adjacent cell, then the profile of the u^+ [−] in the cell according to the wall law may overestimate or underestimate the actual flow profile of u^+ [−]. Assuming the flow velocity, $\langle U \rangle_p$

or $\langle U \rangle_{avg}$ [ms^{-1}], is the same, argue whether the flow friction velocity u_τ [ms^{-1}] as a result of using a wall function will be over- or under-predicted based on whether the profile of u^+ [$-$] is over- or under-estimated. Is your answer the same for point-wise and integrated wall functions? Hint: you may draw the profile of u^+ [$-$] versus y^+ [$-$] and use the definition of u^+ [$-$].

18.4 When implementing the standard wall model for numerical simulations, we should express the Reynolds stress term in the mean momentum equation:

$$\underbrace{\frac{\overline{D}\langle U_j \rangle}{\overline{D}t}}_{\text{Mean Substantial Derivative of Mean}} = \underbrace{\nu \nabla^2 \langle U_j \rangle}_{\text{Surface Forces}} - \underbrace{\frac{\partial \langle u_i u_j \rangle}{\partial x_i}}_{\text{Reynolds Stresses}}$$

$$- \underbrace{\frac{1}{\rho}\frac{\partial \langle p \rangle}{\partial x_j}}_{\text{Normal and Body Forces}} , \qquad (18.23)$$

only in terms of constants and other solution variables, namely average velocity, turbulence kinetic energy, and turbulence kinetic energy dissipation rate. Show that the standard wall model can be used to express Reynolds stress in a form that does not include the nominal friction velocity, u_τ^* [ms^{-1}], so that

$$- \langle uv \rangle_p = \frac{c_\mu^{1/4} k_p^{1/2} \langle U \rangle_p}{\frac{1}{\kappa} \ln \left(\frac{c_\mu^{1/4} k_p^{1/2} y_p}{\nu} \right) + B} . \qquad (18.24)$$

18.5 Various specific wall models are used to describe atmospheric flows. Wall models for the atmosphere describe vertical variation of horizontal wind speed within the first few tens of metres away from the earth surface. For instance, [28] provide the following logarithmic wall model

$$\langle U \rangle (z) = \frac{u_\tau}{\kappa} \ln \left(\frac{z}{z_0} \right) , \qquad (18.25)$$

where $\langle U \rangle (z)$ [ms^{-1}] is horizontal wind speed as a function of height z [m], u_τ [ms^{-1}] is friction velocity, κ [$-$] is the von Kármán constant, and z_0 [m] is the earth surface aerodynamic roughness height. For instance, over grassland it may be assumed $z_0 = 0.1$ m, while in high-rise urban areas it may be assumed $z_0 = 10$ m. Assuming $u_\tau = 1$ ms^{-1}, calculate horizontal wind speed at $z = 100$ m over grassland versus high-rise urban areas. What is the difference between the calculated horizontal wind speeds?

18.6 Some atmospheric boundary layer wall models assume a displacement height d [m], which will shift the wind profile in the z direction with magnitude d [m]. The rationale for this modification is the observation particularly in the densely built urban environment [11]. References [11] and [29] provide one such wall model as

$$\langle U \rangle (z) = \frac{u_\tau}{\kappa} \ln \left(\frac{z - d}{z_0} \right) , \tag{18.26}$$

where $\langle U \rangle (z)$ [ms^{-1}] is horizontal wind speed as a function of height z [m], u_τ [ms^{-1}] is friction velocity, κ [$-$] is the von Kármán constant, and z_0 [m] is the earth surface aerodynamic roughness height. Assuming $z_0 = 10$ m, $u_\tau = 1$ ms^{-1}, and $d = 5$ m calculate horizontal wind speed at $z = 100$ m. What is the difference between this calculation compared to the same calculation in the previous problem?

18.7 Alternative to the logarithmic wall model, a power-law wall model has been proposed for the atmospheric boundary layer. Reference [28] provide this wall model as

$$\langle U \rangle (z) = \langle U \rangle (z_{ref}) \left(\frac{z}{z_{ref}} \right)^\alpha , \tag{18.27}$$

where z_{ref} [m] is some reference height, $\langle U \rangle (z_{ref})$ [ms^{-1}] is horizontal wind speed at this reference height, and α [$-$] is a fitted constant. We wish to express α [$-$] as a function of z_{ref} and z_0 [m] if the logarithmic and power-law wall functions are to be matched at reference height z_{ref} [m]. To do this, follow these steps. First rearrange the power-law wall model to express α [$-$] as

$$\alpha = \frac{\ln \left(\frac{\langle U \rangle (z)}{\langle U \rangle (z_{ref})} \right)}{\ln \left(\frac{z}{z_{ref}} \right)} . \tag{18.28}$$

Next substitute both $\langle U \rangle (z)$ and $\langle U \rangle (z_{ref})$ [ms^{-1}] by their logarithmic wall model expressions. After doing this α [$-$] can be expressed as a function of z, z_{ref}, and z_0 [m]. Next take the limit of

$$\lim_{z \to z_{ref}} \left(\alpha (z, z_{ref}, z_0) \right) . \tag{18.29}$$

This limit can be conveniently taken by invoking the L'Hospital's Rule. After taking this limit, obtain the following relationship for α [$-$]:

$$\alpha = \frac{1}{\ln \left(\frac{z_{ref}}{z_0} \right)} . \tag{18.30}$$

18.8 An oceanographic scientist has developed a Large-Eddy Simulation (LES) model to simulate water flow in the ocean on top of a rough ocean bed. Suppose the characteristic length at the ocean floor is L [m]. For instance, this could be the depth

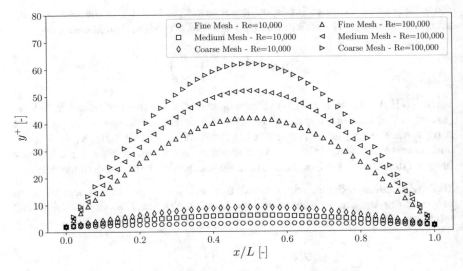

Fig. 18.4 Ocean flow Large-Eddy Simulation (LES): y^+ [−] versus x/L [−]

of an ocean valley or the height of an ocean hill. Flow is in the positive x direction parallel to ocean floor while the direction normal to the ocean floor is positive y. The scientist creates three numerical grids: fine, medium, and coarse. The fine grid has the smallest numerical cells while the coarse grid has the largest numerical cells. The scientist simulates ocean flow at two water velocities that correspond to Reynolds numbers 10,000 and 100,000. In total, the scientist has six simulations. For each simulation, the scientist provides a plot of y^+ [−] against non-dimensional distance x/L [−]. y^+ [−] is calculated using the height of the centre of the first cell adjacent to the ocean bed. This plot is provided in Fig. 18.4.

Provide an argument to explain which of the simulations above require a wall model to provide accurate results. Why does y^+ [−] increase by coarsening the numerical grid? Why does y^+ [−] increase by increasing the flow Reynolds number?

18.9 In atmospheric science, surface layer is defined as a layer of flow adjacent to the earth surface within the boundary layer where the magnitude of the friction velocity is constant [30]. This friction velocity in atmospheric flows is denoted by $u_{\tau,ABL}$ [ms^{-1}], where ABL stands for Atmospheric Boundary Layer (ABL). The height of the surface layer is usually about 10% of the height of ABL. In fact in many CFD analyses applied to ABL, where the height of the computational domain is less than 10% of the height of ABL, it is fair to assume a constant atmospheric friction velocity $u_{\tau,ABL}$ [ms^{-1}] throughout all elevations for upstream flow. With analogy to the material found in this chapter, provide an argument to support the following relationships:

$$\langle U \rangle(y) = \frac{u_{\tau,ABL}}{\kappa} \ln\left(\frac{y}{y_0}\right) , \tag{18.31}$$

$$k(y) = \frac{u_{\tau,ABL}^2}{\sqrt{C_\mu}} , \tag{18.32}$$

$$\epsilon(y) = \frac{u_{\tau,ABL}^3}{\kappa y} , \tag{18.33}$$

where y [m] is vertical distance normal to the earth surface, y_0 [m] is the aerodynamic roughness height of the earth surface, $\langle U(y) \rangle$ [ms^{-1}] is mean horizontal velocity in the x direction, κ [−] is the von Kármán constant, $k(y)$ [m^{2}s^{-2}] is turbulence kinetic energy, C_μ [−] is a model constant, and $\epsilon(y)$ [m^{2}s^{-3}] is turbulence kinetic energy dissipation rate. Comment if the turbulence kinetic energy is or is not a function of height y [m] within the surface layer.

18.10 In this chapter the log-law of the wall was introduced as $u^+ = \frac{1}{\kappa} \ln(y^+) + B$ [−]. This law is valid for rather smooth walls. It has been shown that this model must be modified for rough walls to arrive at a new law that is the log-law for a rough wall [26]. To express this new law first we define the dimensionless physical roughness height such that

$$k_S^+ = \frac{u_\tau k_S}{\nu} , \tag{18.34}$$

where k_S [m] is the roughness characteristic height. It has been reported that the log-law of the wall for a rough wall does not exhibit a viscous or buffer sublayer. In other words, these sublayers are destroyed or eliminated. The new law is simply obtained by shifting the intercept of the curve for log-law [26] such that

$$u^+ = \frac{1}{\kappa} \ln(y^+) + B - \Delta B(k_S^+) , \tag{18.35}$$

where the magnitude of the shift in intercept $\Delta B(k_S^+)$ [−] is a function of k_S^+ [−]. This intercept shift is reported as

$$\Delta B(k_S^+) = \frac{1}{\kappa} \ln(k_S^+) - 3.3 . \tag{18.36}$$

This equation is valid for dimensionless physical roughness heights as large as $k_S^+ = 10,000\ [-]$. Show that if $B = 5.2\ [-]$ then the log-law for a rough wall can be written as

$$u^+ = \frac{1}{\kappa} \ln\left(\frac{y^+}{k_S^+}\right) + 8.5\ . \tag{18.37}$$

18.11 In the previous two problems two wall models have been proposed for the atmospheric boundary layer. These were

$$\langle U \rangle (y) = \frac{u_\tau}{\kappa} \ln\left(\frac{y}{y_0}\right)\ , \tag{18.38}$$

$$u^+ = \frac{1}{\kappa} \ln\left(\frac{y^+}{k_S^+}\right) + 8.5\ . \tag{18.39}$$

If these two wall models were to be matched, then there will be a relationship between k_S and y_0 [m], i.e. k_S [m] will be a multiplier of y_0 [m]. If $\kappa = 0.4\ [-]$, show that this relationship can be approximated by:

$$k_S \approx 30 y_0\ . \tag{18.40}$$

In fact both k_S and y_0 [m] are measures of earth surface roughness characterizing a flat horizontal surface. y_0 [m] is known as aerodynamic roughness height while k_S [m] is known as sand-grain roughness height [26]. From the point of view of meteorology, earth surface is never smooth but rough due to trees, buildings, etc.

18.12 An engineer is designing two types of wall functions for temperature in a boundary layer flow along a surface with direction z normal to the surface. Using Reynolds averaging, vertical velocity normal to the surface can be expressed as $W = \langle W \rangle + w\ [\mathrm{ms}^{-1}]$, where $\langle W \rangle = 0\ \mathrm{ms}^{-1}$, and temperature can be expressed as $T = \langle T \rangle + t\ [\mathrm{K}]$. For the wall, the surface heat flux is specified as q_w in $[\mathrm{Wm}^{-2}]$. This is the amount of heat transferred to the fluid, positive when heat is added to the fluid, normal to the surface. The engineer wishes to use q_w $[\mathrm{Wm}^{-2}]$ and implement (a) wall function 1 for forcing the turbulent kinematic heat flux at point p, i.e. $\langle wt \rangle_p$ $[\mathrm{Kms}^{-1}]$, away from the surface and in the middle of the first computational cell adjacent to the wall and (b) wall function 2 for forcing mean temperature gradient at point p, i.e. $\frac{\partial \langle T \rangle}{\partial z}|_p$ $[\mathrm{Km}^{-1}]$, away from the surface and in the middle of the first computational cell adjacent to the wall. (a) For wall function 1, perform unit analysis to show that the dynamic and kinematic heat fluxes are related in such a way to result in the following simple wall function:

$$\langle wt \rangle_p = \frac{q_w}{\rho C_p}\ , \tag{18.41}$$

where ρ [kgm^{-3}] is fluid density and C_p [Jkg^{-1}K^{-1}] is the heat capacity of the fluid. Argue in which region within the law of the wall for temperature the viscous heat transfer may be ignored so that this wall function is valid. (b) For wall function 2, use the gradient diffusion hypothesis to show that the wall function can be written as

$$\frac{\partial \langle T \rangle}{\partial z}\Big|_p = -\frac{q_w}{\rho C_p \alpha_t} = -\frac{q_w Pr_T}{\rho C_p \nu_t} , \qquad (18.42)$$

where ν_T [m^{2}s^{-1}] is turbulent viscosity and Pr_T [$-$] is turbulent Prandtl number. Again, argue for which region within the law of the wall for temperature the viscous heat transfer may be ignored so that this wall function is valid.

18.13 Using Monin–Obukhov similarity theory, integrate the relationships for momentum, potential temperature, and specific humidity from, z_0, z_T, and z_Q [m], respectively, to $z - d$ [m] to obtain relationships for the three variables as a function of height z [m], i.e. $\langle U \rangle(z)$ [ms^{-1}], $\langle \Theta \rangle(z)$ [K], and $\langle Q \rangle(z)$ [kgkg^{-1}]. Only perform the integration under thermally stable and neutral conditions. For the thermally unstable condition, based on the paper by [20], adapt the derived formulation by [20] for your case to express the three variables $\langle U \rangle(z)$ [ms^{-1}], $\langle \Theta \rangle(z)$ [K], and $\langle Q \rangle(z)$ [kgkg^{-1}] under the thermally unstable case.

18.14 An engineer wishes to extend the Monin–Obukhov similarity theory for a passive scalar $C = \langle C \rangle + c$ [$-$] in the atmospheric surface layer. Suppose that the vertical flux of the passive scalar, i.e. $\langle wc \rangle$ [ms^{-1}], is known. Also suppose that the principle of heat and mass transfer similarity can be used to establish $\Phi_C\left(\frac{z}{L}\right) = \Phi_H\left(\frac{z}{L}\right)$ [$-$], that is the universal functions for the passive scalar and heat are similar. Also, assume that the roughness length for the passive scalar satisfies $z_C = 0.1z_0$ [m]. Formulate a differential equation to relate $\frac{d\langle C \rangle}{dz}$ [m^{-1}] and $\Phi_C\left(\frac{z}{L}\right)$ [$-$]. Integrate this differential equation from z_C to z [m] to obtain a function for $\langle C \rangle(z)$ [$-$] under the thermally stable and neutral cases. For the thermally unstable condition, based on the paper by [20], adapt the derived formulation by [20] for your case to express $\langle C \rangle(z)$ [$-$] under the thermally unstable case.

References

1. Launder B E, Spalding D B (1972) Mathematical models of turbulence. Academic Press, London
2. Werner H, Wengle H (1993) Large-eddy simulation of turbulent flow over and around a cube in a plate channel. In: Durst F, Friedrich R, Launder B E, Schmidt F W, Schumann U, Whitelaw J H (ed) Turbulent Shear Flows 8. Springer, Berlin
3. Efros V (2006) Large eddy simulation of channel flow using wall functions. Chalmers University of Technology, Göteborg
4. Smagorinsky J (1963) General circulation experiments with the primitive experiments: I. The basic equations. Mon Weather Rev 91:99–164

5. van Driest E R (1956) On turbulent flow near a wall. J Aeronaut Sci 23:1007–1011
6. Zilitinkevich S, Baklanov A (2002) Calculation of the height of the stable boundary layer in practical applications. Bound-Lay Meteorol 105:389–409
7. Zilitinkevich S (2002) Third-order transport due to internal waves and non-local turbulence in the stably stratified surface layer. Q J Roy Meteor Soc 128:913–925
8. Aliabadi A A, Staebler R M, de Grandpré J et al (2016) Comparison of estimated atmospheric boundary layer mixing height in the Arctic and southern Great Plains under statically stable conditions: experimental and numerical aspects. Atmos-Ocean 54:60–74
9. Monin A S, Obukhov A M (1954) Basic laws of turbulent mixing in the surface layer of the atmosphere. Contrib Geophys Inst Acad Sci USSR 151:e187
10. Foken T (2006) 50 years of the Monin–Obukhov similarity theory. Bound-Lay Meteorol 119:431–447
11. Raupach M R, Antonia R A, Rajagopalan S (1991) Rough-wall turbulent boundary layers. Appl Mech Rev 44:1–25
12. Brutsaert W (1982) Evaporation into the atmosphere. Springer, Dordrecht
13. Garratt J (1994) The atmospheric boundary layer. Cambridge University Press, Cambridge
14. Järvi L, Grimmond C S B, Christen A (2011) The surface urban energy and water balance scheme (SUEWS): evaluation in Los Angeles and Vancouver. J Hydrol 411:219–237
15. Meili N, Manoli G, Burlando P et al (2020) An urban ecohydrological model to quantify the effect of vegetation on urban climate and hydrology (UT&C v1.0). Geosci Model Dev 13:335–362
16. Businger J A, Wyngaard J C, Izumi Y et al (1971) Flux-profile relationships in the atmospheric surface layer. J Atmos Sci 28:181–189
17. Dyer A J (1974) A review of flux-profile relationships. Bound-Lay Meteorol 7:363–372
18. Zeng X, Dickinson R E (1998) Effect of surface sublayer on surface skin temperature and fluxes. J Climate 11:537–550
19. Moradi M, Dyer B, Nazem A et al (2021) The Vertical City Weather Generator (VCWG v1.3.2). Geosci Model Dev 14:961–984
20. Paulson C A (1970) The mathematical representation of wind speed and temperature profiles in the unstable atmospheric surface layer. J Appl Meteorol 9:857–861
21. Hanna S R, Britter R E (2002) Wind flow and vapor cloud dispersion at industrial and urban sites. American Institute of Chemical Engineers, New York
22. Temmerman L, Leschziner M A, Mellen C P et al (2003) Investigation of wall-function approximations and subgrid-scale models in large eddy simulation of separated flow in a channel with streamwise periodic constrictions. Int J Heat Fluid Fl 24:157–180
23. Raupach M R, Thom A S, Edwards I (1980) A wind-tunnel study of turbulent flow close to regularly arrayed rough surfaces. Bound-Lay Meteorol 18:373–397
24. Raupach M R (1992) Drag and drag partition on rough surfaces. Bound-Lay Meteorol 60:375–395
25. Jiménez J (2004) Turbulent flows over rough walls. Annu Rev Fluid Mech 36:173–196
26. Blocken B, Stathopoulos T, Carmeliet J (2007) CFD simulation of the atmospheric boundary layer: Wall function problems. Atmos Environ 41:238–252
27. Qi M, Li J, Chen Q et al (2018) Roughness effects on near-wall turbulence modelling for open-channel flows. J Hydraul Res 56:648–661
28. Seinfeld J H, Pandis S N (2006) Atmospheric chemistry and physics from air pollution to climate change, 2nd edn. John Wiley & Sons, Inc, Hoboken
29. Graf A, van de Boer A, Moene A et al (2014) Intercomparison of methods for the simultaneous estimation of zero-plane displacement and aerodynamic roughness length from single-level eddy-covariance data. Bound-Lay Meteorol 151:373–387
30. Stull R B (1998) An introduction to boundary layer meteorology. Kluwer Academic Publishers, Dordrecht

Chapter 19
Model Evaluation

Abstract This chapter lays the foundation for assessment of numerical models related to fluid simulations. The concepts of model verification and validation are described. The concept of time and space discretization error estimation is introduced. The role of turbulence model completeness in error estimation is described. The order of convergence is defined. Two convergence criteria are proposed: the Grid Independence Test (GIT) and the Grid Convergence Index (*GCI*). Finally, a few statistical metrics are provided for comparison of the numerical simulation result of a model against some reference solution, which could be obtained from a set of observations, analytical solution, or another model. These are Bias (B) (or Fractional mean Bias (FB)), Root Mean Square Error (RMSE) (or Normalized Mean Square Error (NMSE)), and the Pearson's Correlation Coefficient (PCC).

19.1 Overview

Any turbulence model must be carefully assessed before it could be concluded that it is suitable for a particular application or analysis. Especially after introducing numerous models in this book (Chaps. 15 to 18), it is necessary to consider a variety of tools and techniques in assessing turbulence models. This is the main goal of this chapter.

19.2 Verification and Validation

One of the main pillars of numerical simulation is the validation and verification of the calculations performed using a specific code or technique. *Validation* is the process of solving the right equations for the simulated physics, while *verification* is solving those equations in a proper manner. Based on the two definitions, one cannot validate a whole numerical code, but only a specific set of calculations for a case study performed using the code [1].

19.3   Time and Space Discretization Error Estimation

Discretization convergence is an analysis that must be performed for most *complete models* of turbulence, such as the $k - \epsilon$ model, that are solved numerically. However, although informative, this analysis is not necessarily conclusive when applied to *incomplete models* of turbulence, such as the LES or mixing length models.

Spatial and temporal discretization of a model domain is a crucial component for numerical simulation. The number of time steps or grid elements, or nodes, that can be created in one domain may vary significantly depending on the time march sequence, size of that domain, and the sizes of the grid elements themselves. As the numerical solution of the governing equations is obtained for each time step and each spatial element or node of the grid, the number of time steps, elements, or nodes and the way they are arranged in the grid can notably affect the accuracy of the numerical results.

19.4   Order of Convergence

The *order of grid convergence* involves the behaviour of the solution error defined as the difference between the discrete solution and the exact solution:

$$E = f(h) - f_{exact} = Ch^p + H.O.T. , \tag{19.1}$$

where C is a constant, h [m] is some measure of mesh or grid spacing, and p [−] is the order of convergence. The Higher Order Terms ($H.O.T.$) are negligible compared to Ch^p. A representative cell mesh size h [m] can be defined as

$$h = \left(\frac{1}{N} \sum_{i=1}^{N} \Delta V_i \right)^{1/3} , \tag{19.2}$$

where ΔV_i [m^3] is the volume of cell i and N [−] is the total number of cells. Note that this is defined for *Control Volume Schemes (CVS)*, but for *Finite Difference Schemes (FDS)* or *Finite Element Schemes (FES)*, the representative cell mesh size h [m] can be defined in a similar fashion.

A *second-order discretization* for either space or time means that p [−] is equal, or at least very close, to two. A numerical code uses a numerical algorithm that will provide a theoretical order of convergence; however, the boundary conditions, numerical models, and mesh will reduce this order so that the observed order of convergence will likely be lower. Neglecting $H.O.T$ and taking the logarithm of both sides of the above equation result in

$$\ln E = \ln C + p \ln h . \tag{19.3}$$

The order of convergence p [$-$] can be obtained from the slope of the curve of $\ln E$ versus $\ln h$. If such data points are available, the slope can be read from the graph or the slope can be computed from a least-squares fit to the data.

A more direct evaluation of p [$-$] can be obtained from three solutions. Suppose, we select three significantly different sets of meshes and run our simulations to determine values of key solutions needed for an error estimation study. For example, assume ϕ [$-$] is a solution being reported. We assume $h_1 < h_2 < h_3$ [m]. We define the mesh refinement ratio to be $r_{mn} = h_m/h_n$ [$-$] and further the difference between solutions at two different mesh levels be $\phi_{mn} = \phi_m - \phi_n$ [$-$]. If using a constant mesh refinement ratio, i.e. if $r = r_{32} = r_{21}$ [$-$], then

$$p = \frac{\ln(\phi_{32}/\phi_{21})}{\ln r}. \tag{19.4}$$

The order of convergence is determined by the order of the leading term of the truncation error and is represented with respect to the scale of the discretization, h [m]. The local order of convergence is the order for the stencil representing the discretization of the equation at one location in the mesh, for instance interior or boundary locations. The global order of convergence considers the propagation and accumulation of errors outside the stencil. This propagation causes the global order of convergence to be less than the local order of convergence in the interior of a domain. The order of convergence for the boundary conditions can be one order lower than the interior order of convergence without degrading the overall global order of convergence significantly. This is due to the fact that in three-dimensional geometries boundary cells are outnumbered by interior cells.

Assessing the order of convergence of a code and calculations requires one to sufficiently refine the mesh such that the solution is in the asymptotic range of convergence. The asymptotic range of convergence is obtained when the mesh spacing is such that the various mesh spacings h [m] and errors E result in the constancy of C, i.e.

$$C = \frac{E}{h^p}. \tag{19.5}$$

19.5   Grid Independence Test (GIT)

The most simple test is applied for spatial discretization, where a procedure is performed for quantifying the degree of independence of the numerical solution from the grid size and configuration changes, which is called *Grid Independence Test (GIT)*. The grid independence test is usually performed for three different levels of grid fineness: coarse, medium, and fine. The methodology that is used for grid independence testing consists of a comparison between the obtained solution (velocity, temperature, concentration, etc.) on a continuous spatial segment, such as line or surface, for each grid level. The grid level that exhibits enough grid independency of its solution (shows no significant change in the solution with the change in grid size to a finer level) is chosen for further use. Sometimes grid economy is considered in the independency test. If a finer grid shows no considerable change in the solution than a coarser grid, the latter is *independent enough* and is chosen to accelerate the solution process in later case studies [2].

19.6   Grid Convergence Index (*GCI*)

A representative measure for grid refinement studies was proposed by [3], called *Grid Convergence Index (GCI)* [−]. The GCI [−] is extracted from the generalized Richardson extrapolation theory and uses an asymptotic approach for calculating the amount of uncertainty in grid convergence [3]. Similar to the simpler grid independence test, the GCI [−] makes use of the solution on three different grid size levels, such grids that can be created through grid coarsening and not necessarily by grid refinement [1]. The GCI [−] reports a numeric value that shows how much convergence is achieved in the solution between two successive grid levels, or between the coarsest grid level, taken as a reference, and each one of the two other grids.

A consistent numerical analysis is one which provides a result approaching an asymptotic value as the mesh resolution approaches zero. Thus, the discretized equations will approach the solution of the original differential equations. One significant issue in numerical computations is to decide what level of mesh resolution is appropriate. This is a function of the flow conditions, type of analysis, geometry, and other variables. One is often left to start with a coarse mesh resolution and then conduct a series of mesh refinements to assess the effect of mesh resizing. This is known as a mesh refinement study.

One must recognize the distinction between a numerical result that approaches an asymptotic value and one that approaches the true solution. Even when the asymptotic solution to a set of differential equations is found, it may be different from the true physical solution. The GCI [−] is a measure of the percentage the computed solution is away from the asymptotic computed solution. It indicates an error band on how far the solution is from the asymptotic value and how much the solution would change with a further refinement of the mesh. A small value of GCI [−] indicates that the computation is within the asymptotic range. The GCI [−] is defined as

$$GCI_{mn} = \frac{F_s |\epsilon_{mn}|}{r^p - 1}, \tag{19.6}$$

where F_s [−] is a factor of safety. The refinement may be in either space or time. The factor of safety is recommended to be 3.0 for comparisons of two meshes and

1.25 for comparison over three meshes or more. The relative error ϵ_{mn} [−] is defined by:

$$\epsilon_{mn} = \frac{\phi_m - \phi_n}{\phi_n} . \tag{19.7}$$

It is assumed that the mesh refinement ratio r [−] is applied equally in all coordinate directions (i, j, k) for steady state solutions and also time t [s] for time-dependent solutions. If this is not the case, then the grid convergence indices can be computed for each direction independently and then added to give the overall grid convergence index by:

$$GCI = GCI_t + GCI_x + GCI_y + GCI_z . \tag{19.8}$$

It must be noted that the concept of grid convergence does not always apply to models that are not complete (or incomplete), such as the LES or mixing length models. A model is termed complete if its constituent equations are free from flow-dependent specifications. Such specifications include material properties (density and viscosity), initial and boundary conditions, and numerical discretization. In such a case, a GCI [−] may not necessarily approach zero or even reduce by further refining the mesh, which is the case for models that are redefined at a specific lengthscale. For example, the LES model formulates and solves different sets of partial differential equations at above-grid and subgrid scales [1, 4, 5].

19.7   Reference and Model Error Quantification

Error quantification between the reference quantities (could be obtained from a set of experimentally observed or measured quantities, analytical solution, or another model) and the numerically predicted quantities is performed in various ways. Most commonly, the simple *Bias (B)* or *Root Mean Square Error (RMSE)* are quantified and used to report the amount of shift and spread, respectively, between the observed and predicted values, with the observed quantity usually used as the reference. These errors are defined as

$$B = \langle \phi_p - \phi_o \rangle , \tag{19.9}$$

$$RMSE = \sqrt{\langle (\phi_p - \phi_o)^2 \rangle} , \tag{19.10}$$

where ϕ_p and ϕ_o are the predicted and observed quantities. Although providing absolute measures of error, this method may yield some exaggerated, unrepresentative, and undefined error estimates in some cases. Alternatively, [6] proposed two performance measures to express the error between the observed and predicted quantities: the *Fractional mean Bias (FB)* [−] and the *Normalized Mean Square Error (NMSE)* [−], defined as

$$FB = \frac{2(\langle \phi_p - \phi_o \rangle)}{\langle \phi_p \rangle + \langle \phi_o \rangle} \,, \tag{19.11}$$

$$NMSE = \frac{\langle (\phi_p - \phi_o)^2 \rangle}{\langle \phi_p \rangle \langle \phi_o \rangle} \,. \tag{19.12}$$

Unlike the previous method, FB [−] and $NMSE$ [−] define some specific attributes for the calculated error. Averaged over all data points, the FB [−] represents the shift between the observed and predicted values, while the $NMSE$ [−] gives the spread of one side of the values with respect to the other. For a theoretically perfect model, FB [−] and $NMSE$ [−] should be equal to 0. FB [−] and $NMSE$ [−] can be used with any physical quantity. The caveat in using FB [−] and $NMSE$ [−] is that if either the predicted or observed quantity is near zero, then this method predicts large and exaggerated amounts of error.

Another way to compare a set of predicted and observed quantities is to compute the Pearson's Correlation Coefficient (PCC [−]) defined as

$$PCC = \frac{\langle (\phi_p - \langle \phi_p \rangle)(\phi_o - \langle \phi_o \rangle) \rangle}{\sqrt{\langle (\phi_p - \langle \phi_p \rangle)^2 \rangle} \sqrt{\langle (\phi_o - \langle \phi_o \rangle)^2 \rangle}} \,. \tag{19.13}$$

A PCC [−] close to 1 shows that the predictions are highly correlated with the observed values, while a value close to zero or negative indicates no correlation or anti-correlation, respectively [7].

Problems
19.1 A model error quantification is to be performed against experimental observations. The following set of observations and the corresponding model predictions are provided:

$$\phi_o = 1, 2, 3, 4, 5 \,, \tag{19.14}$$

$$\phi_p = 0, 1, 1, 2, 3 \,. \tag{19.15}$$

Show that the fractional mean bias and the normalized mean square error for the above sets can be calculated as

$$FB = -0.73 \,, \tag{19.16}$$

$$NMSE = 0.67 \,. \tag{19.17}$$

Table 19.1 Various runs of a RANS model given different pairs of cell mesh size h [m] and a solution error E [−]

Run	1	2	3	4	5	6
h [m]	2.72	7.40	20.1	54.6	148	403
E [−]	8.17	44.7	493	3640	19,930	59,874

19.2 Assume that the order of convergence for a discretized model is exactly $p = 2$ [−]. Show that if the resolution of discretization is increased by a factor of two, i.e. $h_2 = \frac{1}{2}h_1$ [m], then the discretization error is reduced by a factor of four, i.e.

$$E_2 = \frac{1}{4}E_1 . \tag{19.18}$$

19.3 Reason why the safety factor $F_s = 3.0$ [−] is larger when calculating GCI [−] for two levels of mesh compared to the safety factor $F_s = 1.25$ [−] used when calculating GCI [−] for tree levels of mesh.

19.4 A marine engineer is performing convergence analysis of a Reynolds-Averaged Navier–Stokes (RANS) turbulence model for an engineering problem involving ocean flow around a massive oil tanker. He has run the model on various representative cell mesh sizes h [m] and obtained various errors E [−] for a particular solution of the model. Table 19.1 shows numerous pairs of h [m] and E [−].

Help him write a Python code to calculate the order of convergence p [−] for this RANS model. This can be achieved by fitting the data. Show your (a) Python code, (b) its console output, and (c) plot of the curve for natural logarithm of error versus natural logarithm of mesh cell size.

19.5 An engineer is calculating the Grid Convergence Index (GCI [−]) for her turbulence model studying the solution ϕ [−] obtained on two levels of grid, a coarse and a fine grid, with a grid refinement ratio of $r = 1.5$ [−]. The model exhibits an order of convergence of $p = 1.8$ [−]. The two successive solutions obtained on each grid result in the solution $\phi_m = 2$ [−] on the coarse grid and solution $\phi_n = 2.1$ [−] on the fine grid. Assuming $F_s = 3.0$ [−], calculate the GCI_{mn} [−].

References

1. Roache P J (1997) Quantification of uncertainty in computational fluid dynamics. Annu Rev Fluid Mech 29:123–160
2. Elmaghraby H A, Chiang Y W, Aliabadi A A (2018) Ventilation strategies and air quality management in passenger aircraft cabins: A review of experimental approaches and numerical simulations. Sci Technol Built En 24:160–175
3. Roache P J (1994) Perspective: A method for uniform reporting of grid refinement studies. J Fluid Eng 116:405–413
4. Poletto R, Craft T, Revell A (2013) A new divergence free synthetic eddy method for the reproduction of inlet flow conditions for LES. Flow Turb Combust 91:519–539

5. Aliabadi A A, Krayenhoff E S, Nazarian N et al (2017) Effects of roof-edge roughness on air temperature and pollutant concentration in urban canyons. Bound-Lay Meteorol 164:249–279
6. Hanna S, Chang J (2012) Acceptance criteria for urban dispersion model evaluation. Meteorol Atmos Phys 116:133–146
7. Nambiar M K, Robe F, Seguin A M et al (2020) Diurnal and seasonal variation of area-fugitive methane advective flux from an open-pit mining facility in northern Canada using WRF. Atmosphere-Basel 11:1227

Part IV
Applications

Chapter 20
Engineering

Abstract This chapter briefly introduces the engineering fields that require the application of the study of turbulence. While not exhaustive, some example fields include liquid–liquid extraction industries, coalescer, waste water treatment, desalination, combustion devices, indoor ventilation, aeronautics, renewable energy, and river engineering.

20.1 Overview

Fluid flow occurs in many fields in the natural and technical environment. Such flows are either laminar or turbulent. Three characteristics of the fluids that are of special importance are viscosity, density, and compressibility. This chapter provides a non-exhaustive list of engineering applications, in which the study of turbulence is relevant.

20.2 Liquid–Liquid Extraction Industries

The liquid–liquid extraction is the process of separating a liquid solution containing more than one liquid component. One typical process is *solvent partitioning*, where compounds having different solubilities can be separated into two different immiscible liquids. Solvent extraction is widely used in various industries such as production of vegetable oils and biodiesel, processing of perfumes, and reprocessing of nuclear fuels.

20.3 Coalescer

A *coalescer* is a device that divides an emulsion into specific components. It is primarily used in oil refining to remove water from hydrocarbon liquids and gases to produce high quality hydrocarbon products. For instance, in natural gas industries, a coalescer is used to recover a lube oil from natural gas at the downstream location of a compressor. In a coalescer, droplets will stay in the streamlines around a wire or fibre target, where they are expected to be collected. Usually laminar flow conditions are required in a coalescer since high fluid velocities overcome surface tension forces and strip droplets out of the coalescer medium. This results in reentrainment in flow and prevents droplets from being collected. Slower velocities result in greater residence time in the media and therefore more time for droplets to be collected. In coalescers, turbulent flow conditions must be prevented.

20.4 Waste Water Treatment

In waste water treatment, the process of mixing is characterized by high flow velocities that create turbulent eddies and disperse components in a fluid. For example, the mixer in the flocculation tank mixes the water and the unwanted particles. The turbulent flow will then slowly change to the laminar flow as it moves towards the sedimentation tank. An understanding of the flow regime, either laminar or turbulent, in a sedimentation tank is also essential to properly predict the settling velocity of components. This leads to proper design of the tank for adequate settling of components [1]. Such processes require a detailed understanding of turbulence.

20.5 Desalination

Desalination is the process of obtaining water with low mineral concentrations from water with high mineral concentrations. A typical application is obtaining fresh water from oceans. Desalinators use a *direct contact membrane*, into which flows a low viscosity liquid in the turbulent regime. Turbulent flow is desired in such membranes to reduce head losses along the membrane [2].

20.6 Combustion Devices

An essential indicator of combustion performance is how well a combustion device can mix the fuel and air so that combustion chemistry can occur at stoichiometric conditions, i.e. the right amount of air and fuel concentrations can be achieved over the entire combustion domain for complete combustion. Turbulent mixing is usually designed to be maximized in combustion devices to ensure this condition. These devices are ubiquitous in automobile engines, aircraft engines or turbines, power plant turbines, combined heat and power systems, furnaces, and boilers [3–8].

20.7 Indoor Ventilation

Ventilation is the intentional introduction of fresh or recirculated air into a space. Ventilation is mainly used to control indoor air quality by diluting and displacing indoor pollutants; it can also be used for purposes of thermal comfort or humidification/dehumidification when the introduction of air will help achieve desired indoor psychrometric conditions. The intentional introduction of air can be categorized as either mechanical ventilation or natural ventilation. Mechanical ventilation uses fans to drive the flow of air into a space. Natural ventilation is the intentional passive flow of air into a space through planned openings. Natural ventilation does not require mechanical systems to move air because it relies entirely on passive physical phenomena, such as diffusion, wind pressure, or the stack effect. Mixed mode ventilation systems use both mechanical and natural processes. Airflow in buildings and enclosed spaces, such as transportation devices, typically occurs in the turbulent regime, so a thorough understanding of turbulent flow is necessary for proper design and maintenance of ventilation systems [9–13].

20.8 Aeronautics

Aircraft fly in the turbulent atmosphere at high speeds. In addition, aircraft themselves create wake turbulence behind them as they move through the atmosphere. The presence of turbulent air around an aircraft has implications in lift and drag forces and ultimately in the navigability of the aircraft. The design and operation of aircraft require a full understanding of turbulence so that they can be built economically and operated safely. Aircraft flying at a considerable fraction of the speed of sound enter a flow regime called *compressible flow*. In this regime density variations of air must be considered around the aircraft and the transport equations must be altered and solved accordingly to allow for this important feature of the flow [14].

20.9 Renewable Energy

Many renewable energy conversion devices involve fluid flows, particularly in the turbulent regime. Some examples include wind turbines [15], hydro power turbines, Archimedes screws [16, 17], and bio-oil combustion devices [5, 6]. Again, an understanding of turbulent flows will help designing, operating, and maintaining such devices efficiently.

20.10   River Engineering

River engineering involves human intervention in the course, characteristics, or flow of a river with the intention to produce benefits for humans. Some examples include water resource management, flood protection, and hydropower. Water flow in rivers is usually in the turbulent regime and involves detailed understanding of turbulent transport processes. Rivers are also studied for their sedimentation and erosion behaviour, which are processes that are influenced significantly by turbulent transport [18, 19].

Problems

20.1 Would you expect the Reynolds number of flows around an airplane be higher or the Reynolds number of flows around ships? Why?

20.2 Speculate why turbulence inside internal combustion engines, where air and fuel mix before burning, is desired.

References

1. Gao H, Stenstrom M K (2018) Evaluation of three turbulence models in predicting the steady state hydrodynamics of a secondary sedimentation tank. Water Res 143:445–456
2. Sharqawy M H, Lienhard V J H, Zubair S M (2010) Thermophysical properties of seawater: a review of existing correlations and data. Desalin Water Treat 16:354–380
3. Heywood J B (1988) Internal combustion engine fundamentals. McGraw-Hill Inc., New York
4. Aliabadi A A, Wallace J S (2009) Cost-effective and reliable design of a solar thermal power plant. T Can Soc Mech Eng 33:25–37
5. Aliabadi A A, Thomson M J, Wallace J S et al (2009) Efficiency and emissions measurement of a Stirling-engine-based residential microcogeneration system run on diesel and biodiesel. Energ Fuel 23:1032–1039
6. Aliabadi A A, Thomson M J, Wallace J S (2010) Efficiency analysis of natural gas residential micro-cogeneration systems. Energ Fuel 24:1704–1710
7. Salehi M M, Bushe W K (2010) Presumed PDF modeling for RANS simulation of turbulent premixed flames. Combust Theor Model 14:381–403
8. Salehi M M, Bushe W K, Daun K J (2012) Application of the conditional source-term estimation model for turbulence–chemistry interactions in a premixed flame. Combust Theor Model 16:301–320
9. Aliabadi A A, Rogak S N, Bartlett K H et al (2011) Preventing airborne disease transmission: review of methods for ventilation design in health care facilities. Advances in Preventive Medicine 2011:124064
10. Elmaghraby H A, Chiang Y W, Aliabadi A A (2018) Ventilation strategies and air quality management in passenger aircraft cabins: A review of experimental approaches and numerical simulations. Sci Technol Built En 24:160–175
11. Elmaghraby H A, Chiang Y W, Aliabadi A A (2019) Are aircraft acceleration-induced body forces effective on contaminant dispersion in passenger aircraft cabins? Sci Technol Built En 25:858–872
12. Elmaghraby H A, Chiang Y W, Aliabadi A A (2020) Airflow design and source control strategies for reducing airborne contaminant exposure in passenger aircraft cabins during the climb leg. Sci Technol Built En 26:901–923

13. Elmaghraby H A, Chiang Y W, Aliabadi A A (2020) Normal and extreme aircraft accelerations and the effects on exposure to expiratory airborne contaminant inside commercial aircraft cabins. Sci Technol Built En 26:924–927
14. Schlichting H (1979) Boundary-layer theory, 7th edn. McGraw-Hill, New York
15. Hong J, Toloui M, Chamorro L P et al (2014) Natural snowfall reveals large-scale flow structures in the wake of a 2.5-MW wind turbine. Nat Commun 5:4216
16. Lyons M, Simmons S, Fisher M et al (2020) Experimental investigation of Archimedes screw pump. J Hydraul Eng 146:04020057
17. Simmons S, Dellinger G, Lyons M et al (2021) Effects of inclination angle on archimedes screw generator power production with constant head. J Hydraul Eng 147:04021001
18. Cheng W, Fang H, Lai H et al (2018) Effects of biofilm on turbulence characteristics and the transport of fine sediment. J Soil Sediment 18:3055–3069
19. He C, Nguyen D (2019) Erodibility study of sediment in a fast-flowing river. Int J Sediment Res 34:144–154

Chapter 21
Sciences

Abstract This chapter briefly introduces fields of science that require the application of the study of turbulence. On the broader scales, these fields include meteorology, oceanography, and space.

21.1 Overview

Turbulence is the dominating force in nature from the earth to the outer space. Our understanding of turbulent mechanisms in the nature has been improving by continuous observations and study of the relevant processes through the scientific efforts. This chapter provides a non-exhaustive list of fields in sciences, in which the study of turbulence is relevant.

21.2 Meteorology

Atmospheric turbulences are small-scale and irregular air motions characterized by winds that vary in speed and direction. Turbulence is important because it mixes and churns the atmosphere and causes water vapour, smoke, and other substances, as well as energy, to become distributed both vertically and horizontally.

Atmospheric turbulence near the earth's surface differs from that at higher levels. At low levels (within a few hundred metres of the surface), turbulence has a marked diurnal variation under partly cloudy and sunny skies, reaching a maximum about midday. This occurs because, when solar radiation heats the surface, the air above it becomes warmer and more buoyant, and cooler, denser air descends to displace it. The resulting vertical movement of air, together with flow disturbances around surface obstacles, makes low-level winds extremely irregular. At night the surface cools rapidly, chilling the air near the ground; when that air becomes cooler than the air above it, a stable temperature inversion is created, and wind speed and gustiness both decrease sharply. When the sky is overcast, low-level air temperatures vary much less between day and night, and turbulence remains nearly constant [1].

Several studies have investigated atmospheric turbulence in urban [2–4] and remote areas [5–7].

At altitudes of several thousand metres or more, frictional effects of surface topography on the wind are greatly reduced, and the small-scale turbulence characteristic of the lower atmosphere is absent. Although upper-level winds are usually relatively regular, they sometimes become turbulent enough to affect aviation.

21.3   Oceanography

The ocean circulation is turbulent in the sense that motions on a wide range of scales from a few centimetres to thousands of kilometres continuously interact. In order to develop theories of the large-scale circulation, which affects our climate, we need to understand these interactions. Researchers tackle motions on all of these scales using a combination of observation, turbulence theories, and high-resolution numerical models [8, 9]. One of the major challenges is to understand how energy is transferred from the thousands of kilometres scales, where oceanic motions are forced by the large-scale atmospheric winds, heat, and freshwater fluxes, to the centimetres scales, where energy is dissipated as heat. The first major transfer is from the large-scale currents to the mesoscale eddies. The large-scale ocean currents are unstable to baroclinic instability, which generate eddies with scales of ten to one hundred kilometres, the mesoscales. The mesoscale eddies then interact and generate submesoscale turbulent filaments on scales from ten kilometres to one hundred metres. These motions are primarily horizontally constrained by the ocean stratification and rotation. Only at scales below approximately one hundred metres, the turbulence becomes three-dimensional and is described as stratified microscale turbulence.

The ocean circulation is dominated by geostrophic eddies, i.e. cyclones and anticyclones with radii of ten to one hundred kilometres. These eddies are the ocean equivalent of the storms we experience in the atmosphere as weather. Eddies play an important role in the transport of heat, carbon, and other climatically important tracers across the oceans. Researchers develop theories for the physics of ocean eddies, their role in climate, and their representation in numerical models used for climate studies. A major outcome of these studies is that lateral mixing by mesoscale eddies is suppressed across strong currents, while it is strong on the flanks of the currents. Researchers have also shown that these variations in lateral mixing are crucial to quantify the impact of mesoscale eddies on the large-scale ocean circulation and climate.

Submesoscale flows are a convoluted web of fronts, with horizontal scales between ten kilometres and one hundred metres, that separate waters of different temperatures and salinities. The importance of submesoscale fronts has increasingly come into focus as an important component of ocean dynamics. The submesoscale fronts are most energetic close to the ocean surface. Just as alveoli facilitate rapid exchange of gases in the lungs, fronts are the ducts through which heat, carbon, oxygen, and other climatically important tracers enter into the deep ocean. Researchers have shown that the submesoscale fronts are generated through instabilities of mesoscale currents confined to the surface mixed layer. The instabilities are stronger in winter, and thus submesoscale flows are stronger in winter than in summer.

Oceanic motions with horizontal scales larger than one kilometre and vertical scales larger than one hundred metres are constrained to flow along density surfaces by the earth's rotation and the density stratification. However at smaller scales, the so-called microscales, these constraints become weak and turbulent motions cross density surfaces. In the upper ocean, microscale turbulence is generated by surface winds, air–sea cooling, or evaporation. In the ocean interior, microscale turbulence develops when internal waves develop strong shears and overturn and break, much like surface gravity waves. These breaking events play a fundamental role in the ocean circulation, because they mix the densest waters at the ocean bottom with the lighter waters above, thereby allowing the densest waters to come back to the surface. Researchers have shown that most of these transformations happen along deep ocean boundaries and not in the ocean interior, as it was previously believed.

21.4 Space

Space weather is driven by the sun, which has a continually changing magnetic field. When this field accumulates excess energy, it can erupt and send energetic particles into space, sometimes towards the earth. Scientists believe that such violent emissions of charged particles, known as coronal mass ejections, are in part fuelled by gusty space winds, or turbulence. Scientists have known of the existence of space turbulence but did not really understand its properties, or how it works. Researchers recently measured space turbulence directly for the first time in the laboratory and finally were able to see the dynamics behind it.

Space turbulence is not the same as the earth's turbulence, the sensation people experience when they are flying or sailing. Unlike wind gusts that create air turbulence on the earth, it turns out that space turbulence—also known as plasma turbulence—results from the interaction of the so-called Alfvén waves, often described as resembling waves that run along a stretched piece of string. It is critically important to understand coronal mass ejections because they have the potential to cause harmful and dramatic effects if they strike the earth's own magnetic field. They could cripple space and satellite systems, GPS, military satellites, and the power grid, among other things.

The surface of the sun is about 6000 degrees Celsius, but if you go into the solar corona, it is about 1 million degrees. We do not understand why that is—what leads to this very hot plasma in the solar corona. The theory is that turbulence plays a very important role. And it is those high temperatures that are responsible for the generation of solar wind. The entire sun is a turbulent ball of plasma. Activity on the sun can generate big blobs of plasma launched from the sun that will collide with the earth's magnetic field, resulting in some serious consequences. About 500 magnetic storms occur during a typical eleven-year solar cycle and already have caused damage. In March 1989, for example, a solar storm caused the entire Quebec power grid to collapse in less than two minutes, affecting six million people in the middle of a Canadian winter. A 2003 Halloween storm prompted a massive blackout in the Northeast, and extensive satellite problems, including the loss of the $450 million Midori-2 research satellite.

The worst solar storm in history occurred earlier, in 1859, when a magnetic explosion on the sun, known as the Carrington Event, after the British astronomer who identified it, prompted a worldwide breakdown in communication systems. In those days, however, the major casualties were to telegraphs—primitive compared to the elaborate electronics currently in use. An event of that magnitude today could prove disastrous. A doomsday scenario could knock out power for months. Imagine life without electricity for months. It would change life as we know it. It is very real, and it could happen. It has only happened once before in history, and that was before we had this extensive power grid. Now the world is very different today and ramifications of such events are different too.

Understanding how space turbulence works may tell us how these coronal ejections occur in the first place, and may give us enough information to predict when one is coming and protect us against it. To see how Alfvén waves produce turbulence, the researchers took a plasma cylinder, eighteen metres long and one metre in diameter, and recreated Alfvén waves in the laboratory, launching two along the cylinder—one from the top going down, and one from the bottom going up. When those two waves crossed through each other, they generated a third Alfvén wave. Scientists were able to measure it and determined that it had all the properties we expected it to have. It was shown that the interaction of two Alfvén waves will create a third. This is the fundamental building block of space turbulence, which consists of thousands of these building blocks. Only with a combination of experimental, theoretical, and numerical efforts can we hope to understand the nature of turbulence in space, with the potential to solve some of the long-standing mysteries of the solar system and the universe.

Problems

21.1 Provide an argument to justify whether atmospheric flows or oceanic flows exhibit higher Reynolds numbers? why?

21.2 Can you provide an example of a turbulent flow within living systems in which the fluid of concern is neither air nor water?

References

1. Stull R B (1988) An introduction to boundary layer meteorology. Kluwer Academic Publishers, Dordrecht
2. Aliabadi A A, Krayenhoff E S, Nazarian N et al (2017) Effects of roof-edge roughness on air temperature and pollutant concentration in urban canyons. Bound-Lay Meteorol 164:249–279

3. Aliabadi A A, Moradi M, Clement D et al (2019) Flow and temperature dynamics in an urban canyon under a comprehensive set of wind directions, wind speeds, and thermal stability conditions. Environ Fluid Mech 19:81–109
4. Aliabadi A A, Moradi M, Byerlay R A E (2021) The budgets of turbulence kinetic energy and heat in the urban roughness sublayer. Environ Fluid Mech 21:843–884
5. Wyngaard J C (2004) Toward numerical modeling in the "Terra Incognita". J Atmos Sci 61:1816–1826
6. Flores F, Garreaud R, Muñoz R C (2014) OpenFOAM applied to the CFD simulation of turbulent buoyant atmospheric flows and pollutant dispersion inside large open pit mines under intense insolation. Comput Fluids 90:72–87
7. Kia S, Flesch T K, Freeman B S et al (2021) Atmospheric transport over open-pit mines: The effects of thermal stability and mine depth. J Wind Eng Ind Aerodyn 214:104677
8. Bennett A (2006) Lagrangian fluid dynamics. Cambridge University Press, Cambridge
9. Badin G, Crisciani F (2018) Variational formulation of fluid and geophysical fluid dynamics. Springer, Cham

Part V
Fundamental Analysis Tools and Principles

Chapter 22
Statistics

Abstract This chapter briefly introduces basic topics in statistics that assist the study of turbulence. The topics involve random variables, event, probability, cumulative distribution function, probability density function, mean and moments, and probability distributions.

22.1 Random Variables

In probability and statistics, a *random variable* is a variable quantity whose value depends on possible outcomes. As a function, a random variable is required to be measurable. For example, a velocity component in turbulent flow U [ms^{-1}] is a random variable [1, 2].

22.2 Event

In probability theory, an *event* is a set of outcomes of an experiment (a subset of the sample spaces), to which a probability is assigned. A single outcome may be an element of many different events, and different events in an experiment are usually not equally likely, since they may include very different groups of outcomes. For example, a set of velocity components in turbulent flow that are all less than a particular value constitute an event $A \equiv \{U < 10\,\mathrm{ms}^{-1}\}$ [1, 2].

22.3 Probability

Probability is the measure of the likelihood that an event will occur. Probability is quantified as a number between 0 and 1, where, loosely speaking, 0 indicates impossibility and 1 indicates certainty. The higher the probability of an event, the

A. A. Aliabadi, *Turbulence*, Mechanical Engineering Series,
https://doi.org/10.1007/978-3-030-95411-6_22

more certain that the event will occur. For instance the probability of event $A \equiv \{U < 10\,\text{ms}^{-1}\}$ is

$$p = P(A) = P\{U < 10\,\text{ms}^{-1}\} . \tag{22.1}$$

The probability of any event always satisfies $0 \le p \le 1$ [−]. In other words, a probability can never be less than zero or greater than 1 [−] [1, 2].

22.4   Cumulative Distribution Function

The probability of any event can be determined from the *cumulative distribution function (CDF)* [−] defined by

$$F(V) \equiv P\{U < V\} , \tag{22.2}$$

where U [−] is the random variable and V [−] is one outcome from the sample space. For example, for two events $B \equiv \{U < V_b\}$ [−] and $C \equiv \{V_a < U < V_b\}$ [−] the probabilities $P(B)$ and $P(C)$ [−] can be calculated as

$$P(B) = P\{U < V_b\} = F(V_b) , \tag{22.3}$$

$$P(C) = P\{V_a < U < V_b\} = P\{U < V_b\} - P\{U < V_a\} = F(V_b) - F(V_a) . \tag{22.4}$$

The three basic properties of CDF [−] are as follows:

$$F(V \to -\infty) = 0 , \tag{22.5}$$

$$F(V \to \infty) = 1 , \tag{22.6}$$

$$V_b > V_a \to F(V_b) > F(V_a) . \tag{22.7}$$

22.5   Probability Density Function

The *probability density function (PDF)* [−] is defined as the derivative of CDF [−],

$$f(V) \equiv \frac{dF(V)}{dV} . \tag{22.8}$$

The PDF [−] has the following four properties:

$$f(V) \geq 0 ,$$
(22.9)

$$P\{V \leq U \leq V + dV\} = F(V + dV) - F(V) = f(V)dV ,$$
(22.10)

$$P\{V_a \leq U \leq V_b\} = F(V_b) - F(V_a) = \int_{V_a}^{V_b} f(V)dV ,$$
(22.11)

$$\int_{-\infty}^{\infty} f(V)dV = 1 .$$
(22.12)

22.6 Mean and Moments

If U [−] is a random variable given with PDF $f(V)$ [−], then the *mean* or *expectation* of U [−] is given by

$$\langle U \rangle \equiv \int_{-\infty}^{\infty} V f(V)dV .$$
(22.13)

The moments of random variable U [−] describe the shape of its PDF $f(V)$ [−]. The nth central moment of U [−] is defined by

$$\mu_n \equiv \int_{-\infty}^{\infty} (V - \langle U \rangle)^n f(V)dV .$$
(22.14)

Note that the zeroth moment of U [−] is equal to 1 ($\mu_0 = 1$) [−], while the first moment of U [−] is equal to zero ($\mu_1 = 0$) [−]. These are evident from evaluating the moment integral [3].

22.7 Probability Distributions

If U [−] is uniformly distributed in the interval $a \leq V < b$ [−], then the PDF [−] of *uniform distribution* for U [−], $f(V)$ [−] is

$$\begin{cases} \frac{1}{b-a} & \text{if } a \leq V < b , \\ 0 & \text{if } V < a \text{ or } V \geq b . \end{cases}$$

If U [−] is exponentially distributed with parameter λ [−], then the PDF of *exponential distribution* for U [−], $f(V)$ [−] is

$$\begin{cases} \frac{1}{\lambda}\exp(-V/\lambda) & \text{if } V \geq 0 \,, \\ 0 & \text{if } V < 0 \,. \end{cases}$$

Of fundamental importance in probability theory and theory of turbulent flows is the *Gaussian distribution* or the *normal distribution* with mean μ [−] and width parameter σ [−]. The PDF [−] for this distribution is given by

$$f(V) = \mathcal{N}(V; \mu, \sigma^2) \equiv \frac{1}{\sigma\sqrt{2\pi}}\exp\left[-\frac{1}{2}(V-\mu)^2/\sigma^2\right]. \tag{22.15}$$

Various other probability distributions can be found in the literature [1, 2].

Problems

22.1 Consider that a random variable U [−] is exponentially distributed with parameter λ [−], for which the PDF [−] for U [−], $f(V)$ [−] is

$$\begin{cases} \frac{1}{\lambda}\exp(-V/\lambda) & \text{if } V \geq 0 \,, \\ 0 & \text{if } V < 0 \,. \end{cases}$$

Derive an expression for the cumulative distribution function (CDF) [−] for U [−], or $F(V)$ [−], as a function of V [−].

22.2 Compare the number of parameters needed to define the PDFs [−] for a uniform, exponential, and Gaussian distributions. Which distribution(s) requires the least number of parameters? Which distribution(s) requires the most number of parameters?

References

1. Barlow R J (1993) Statistics: a guide to the use of statistical methods in the physical sciences, 1st edn. Wiley, New York
2. Walpole R E, Myers, R H, Myers S L et al (2002) Probability & statistics for engineers & scientists, 7th edn. Prentice Hall, Upper Saddle River
3. Pope S B (2000) Turbulent flows. Cambridge University Press, Cambridge

Chapter 23
Mathematics

Abstract This chapter briefly introduces basic topics in mathematics, specifically multi-variable calculus, that assist the study of turbulence. The topics included are the Einstein's notation, Kronecker delta, alternating symbol, position vector, divergence, gradient, curl, Laplacian, dot product of two vectors, cross product of two vectors, material or substantial derivative, and tensors.

23.1 Einstein's Notation

In mathematics, especially in the applications of linear algebra to physics, the Einstein's notation or Einstein summation convention is a notational convention that implies summation over a set of indexed terms in a formula, thus achieving notational brevity. According to this convention, when an index variable appears twice in a single term and is not otherwise defined, it implies summation of that term over all the values of the index. In fluid mechanics the indices can range over the set of $\{1, 2, 3\}$ for three-dimensional space. For instance the expression below can be expanded as the summation

$$U_i \frac{\partial}{\partial x_i} = U_1 \frac{\partial}{\partial x_1} + U_2 \frac{\partial}{\partial x_2} + U_3 \frac{\partial}{\partial x_3} \, , \tag{23.1}$$

where i represents the index and the summation is carried out over a set of three, i.e. for three-dimensional space. In the more familiar notation for Cartesian coordinates, $x_1 = x$, $x_2 = y$, and $x_3 = z$ [−] are the coordinate axes and $U_1 = U$, $U_2 = V$, and $U_3 = W$ [ms^{-1}] are the fluid velocities along the coordinate axes [1–3].

23.2   Kronecker Delta

In mathematics, the Kronecker delta, δ_{ij} [−], is a function of two variables, usually just positive integers. The function is 1 if the variables are equal, and 0 otherwise:

$$\begin{cases} \delta_{ij} = 1 & \text{if } i = j\,, \\ \delta_{ij} = 0 & \text{if } i \neq j\,. \end{cases}$$

This function is named after Leopold Kronecker (1823–1891).

23.3   Alternating Symbol

In mathematics, the alternating symbol, or the Levi-Civita symbol, is a function of three variables, usually positive integers 1, 2, and 3. The function value is 1 if the ordering of the variables is cyclic and −1 if the ordering of the variables is anti-cyclic,

$$\begin{cases} \epsilon_{ijk} = 1 & \text{if } (i, j, k) \text{ are cyclic}\,, \\ \epsilon_{ijk} = -1 & \text{if } (i, j, k) \text{ are anti-cyclic}\,, \\ \epsilon_{ijk} = 0 & \text{otherwise}\,. \end{cases}$$

Cyclic orderings are 123, 231, and 312, while anti-cyclic orderings are 321, 132, and 213; otherwise two or more of the suffixes are the same. The alternating symbol, or Levi-Civita symbol, can be remembered using the visual aid in Fig. 23.1.

23.4   Position Vector

In Cartesian coordinates, the position vector $\mathbf{x}$ [m] gives the position of a general point P [−] using three coordinates relative to the origin O [−]. It is given by

$$\mathbf{x} = x_i \mathbf{e}_i = x_1 \mathbf{e}_1 + x_2 \mathbf{e}_2 + x_3 \mathbf{e}_3\,. \tag{23.2}$$

Fig. 23.1 Visual aid for remembering the Levi-Civita symbol

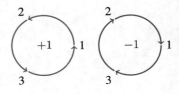

In the more familiar notation for Cartesian coordinates, $x_1 = x$, $x_2 = y$, and $x_3 = z$ are the coordinate axes and $\mathbf{e}_1 = \mathbf{i}$, $\mathbf{e}_2 = \mathbf{j}$, and $\mathbf{e}_3 = \mathbf{k}$ are the unit vectors along the coordinate axes.

23.5 Divergence

In vector calculus, divergence is a vector operator that produces a signed scalar field giving the quantity of a vector field's source at each point. More technically, the divergence represents the volume density of the outward flux of a vector field from an infinitesimal volume around a given point. In physical terms, the divergence of a three-dimensional vector field is the extent to which the vector field flow behaves like a source at a given point. It is a local measure of its *outgoingness*, i.e. the extent to which there is more of some quantity exiting an infinitesimal region of space than entering it. If the divergence is non-zero at some point, then there must be a source or sink at that position. In Cartesian coordinates, if $\mathbf{U}$ [ms^{-1}] is the flow's velocity vector, then divergence is defined as

$$\nabla . \mathbf{U} = \frac{\partial U_i}{\partial x_i} = \frac{\partial U_1}{\partial x_1} + \frac{\partial U_2}{\partial x_2} + \frac{\partial U_3}{\partial x_3} . \tag{23.3}$$

In the more familiar notation for Cartesian coordinates, $x_1 = x$, $x_2 = y$, and $x_3 = z$ are the coordinate axes and $U_1 = U$, $U_2 = V$, and $U_3 = W$ [ms^{-1}] are the fluid velocities along the coordinate axes [1–3].

23.6 Gradient

In mathematics, the gradient is a multi-variable generalization of the derivative. While a derivative can be defined on functions of a single variable, for functions of several variables, the gradient takes its place. The gradient is a vector-valued function, as opposed to a derivative, which is scalar-valued. If f [−] is a differentiable and real-valued function of several variables, its gradient is the vector whose components are the partial derivatives of f [−]. Like the derivative, the gradient represents the slope of the tangent of the graph of the function. More precisely, the gradient points in the direction of the greatest rate of increase of the function, and its magnitude is the slope of the graph in that direction. The gradient (or gradient vector field) of a scalar function f [−] is given by

$$\nabla f = \frac{\partial f}{\partial x_i} \mathbf{e}_i = \frac{\partial f}{\partial x_1} \mathbf{e}_1 + \frac{\partial f}{\partial x_2} \mathbf{e}_2 + \frac{\partial f}{\partial x_3} \mathbf{e}_3 . \tag{23.4}$$

In the more familiar notation for Cartesian coordinates, $x_1 = x$, $x_2 = y$, and $x_3 = z$ are the coordinate axes and $\mathbf{e}_1 = \mathbf{i}$, $\mathbf{e}_2 = \mathbf{j}$, and $\mathbf{e}_3 = \mathbf{k}$ are the unit vectors along the coordinate axes [1–3].

23.7   Curl

In vector calculus, the curl is a vector operator that describes the infinitesimal rotation of a three-dimensional vector field. At every point in the field, the curl of that point is represented by a vector. The attributes of this vector (length and direction) characterize the rotation at that point. The direction of the curl is the axis of rotation, as determined by the right hand rule, and the magnitude of the curl is the magnitude of rotation. If the vector field represents the flow velocity of a moving fluid, then the curl is the circulation density of the fluid. A vector field whose curl is zero is called irrotational. The curl is a form of differentiation for vector fields. The curl of the velocity vector $\mathbf{U}$ [ms^{-1}] is given by

$$\nabla \times \mathbf{U} = \begin{vmatrix} \mathbf{e}_1 & \mathbf{e}_2 & \mathbf{e}_3 \\ \frac{\partial}{\partial x_1} & \frac{\partial}{\partial x_2} & \frac{\partial}{\partial x_3} \\ U_1 & U_2 & U_3 \end{vmatrix}$$

$$\nabla \times \mathbf{U} = \left(\frac{\partial U_3}{\partial x_2} - \frac{\partial U_2}{\partial x_3} \right) \mathbf{e}_1 + \left(\frac{\partial U_1}{\partial x_3} - \frac{\partial U_3}{\partial x_1} \right) \mathbf{e}_2 + \left(\frac{\partial U_2}{\partial x_1} - \frac{\partial U_1}{\partial x_2} \right) \mathbf{e}_3 . \quad (23.5)$$

In the more familiar notation for Cartesian coordinates, where $x_1 = x$, $x_2 = y$, $x_3 = z$, $U_1 = U$, $U_2 = V$, and $U_3 = W$ [ms^{-1}], the curl of velocity vector is given by

$$\nabla \times \mathbf{U} = \begin{vmatrix} \mathbf{i} & \mathbf{j} & \mathbf{k} \\ \frac{\partial}{\partial x} & \frac{\partial}{\partial y} & \frac{\partial}{\partial z} \\ U & V & W \end{vmatrix}$$

$$\nabla \times \mathbf{U} = \left(\frac{\partial W}{\partial y} - \frac{\partial V}{\partial z} \right) \mathbf{i} + \left(\frac{\partial U}{\partial z} - \frac{\partial W}{\partial x} \right) \mathbf{j} + \left(\frac{\partial V}{\partial x} - \frac{\partial U}{\partial y} \right) \mathbf{k} . \quad (23.6)$$

23.8   Laplacian

In mathematics, the Laplace operator or Laplacian is a differential operator given by the divergence of the gradient of a function. In a Cartesian coordinate system, the Laplacian is given by the sum of second partial derivatives of the function with respect to each independent variable. The Laplace operator is named after the

French mathematician Pierre-Simon de Laplace (1749–1827), who first applied the operator to the study of celestial mechanics. The Laplacian occurs in differential equations that describe many physical phenomena, such as electric and gravitational potentials, the diffusion equation for heat and fluid flow, wave propagation, and quantum mechanics. The Laplacian represents the flux density of the gradient flow of a function. For instance, the net rate, at which a chemical dissolved in a fluid moves towards or away from some point, is proportional to the Laplacian of the chemical concentration at that point. The Laplacian is given by

$$\nabla^2 = \nabla.\nabla = \left(e_i \frac{\partial}{\partial x_i}\right).\left(e_j \frac{\partial}{\partial x_j}\right) = \frac{\partial^2}{\partial x_i \partial x_i} . \tag{23.7}$$

For example the Laplacian of function f is given by

$$\nabla^2 f = \frac{\partial^2 f}{\partial x_1^2} + \frac{\partial^2 f}{\partial x_2^2} + \frac{\partial^2 f}{\partial x_3^2} . \tag{23.8}$$

In the more familiar notation for Cartesian coordinates, $x_1 = x$, $x_2 = y$, and $x_3 = z$ [1–3].

23.9 Dot Product of Two Vectors

In mathematics, the dot product or scalar product is an algebraic operation that takes two equal-length sequences of numbers (usually coordinate vectors) and returns a single number. Sometimes it is called inner product in the context of Euclidean space, or rarely projection product for emphasizing the geometric significance. Algebraically, the dot product is the sum of the products of the corresponding entries of the two sequences of numbers. Geometrically, it is the product of the Euclidean magnitudes of the two vectors and the cosine of the angle between them. If u and v [−] represent two vectors expressed in the Cartesian coordinate system, the dot product of the two is given by

$$\mathbf{u}.\mathbf{v} = (e_i u_i).(e_j v_j) = u_i v_i = u_1 v_1 + u_2 v_2 + u_3 v_3 . \tag{23.9}$$

23.10 Cross Product of Two Vectors

In mathematics and vector algebra, the cross product or vector product (occasionally directed area product to emphasize the geometric significance) is a binary operation on two vectors in three-dimensional space. Given two linearly independent vectors, the cross product is a vector that is perpendicular to both vectors and therefore normal to the plane containing them. Because the magnitude of the cross product goes by the sine of the angle between its arguments, the cross product can be thought of as a measure of perpendicularity in the same way that the dot product is a measure of parallelism. If $\mathbf{u}$ and $\mathbf{v}$ [−] represent two vectors expressed in the Cartesian coordinate system, the cross product of the two is given by

$$\mathbf{u} \times \mathbf{v} = \begin{vmatrix} \mathbf{e}_1 & \mathbf{e}_2 & \mathbf{e}_3 \\ u_1 & u_2 & u_3 \\ v_1 & v_2 & v_3 \end{vmatrix} = \epsilon_{ijk} u_j v_k \mathbf{e}_i$$

$$\mathbf{u} \times \mathbf{v} = (u_2 v_3 - u_3 v_2)\mathbf{e}_1 + (u_3 v_1 - u_1 v_3)\mathbf{e}_2 + (u_1 v_2 - u_2 v_1)\mathbf{e}_3 \ . \tag{23.10}$$

Remarkably, the alternating symbol expresses the cross product in a very concise manner. Component i of the cross product is given by $\epsilon_{ijk} u_j v_k$ [−]. For instance the first component of the cross product can be calculated by

$$\epsilon_{123} u_2 v_3 + \epsilon_{132} u_3 v_2 = u_2 v_3 - u_3 v_2 \ . \tag{23.11}$$

23.11 Material or Substantial Derivative

Material derivative or substantial derivative describes the time rate of change of some physical quantity of a material element that is subjected to a space and time dependent macroscopic velocity field variations of that physical quantity. In fluid dynamics, the velocity field is the flow velocity, and the quantity of interest may be the density, momentum, or temperature of the fluid. In this case, the total derivative describes the density, momentum, or temperature change of a certain fluid parcel with time, as it flows along its path line or trajectory. The total derivative is defined as

$$\frac{D}{Dt} \equiv \frac{\partial}{\partial t} + U_i \frac{\partial}{\partial x_i} = \frac{\partial}{\partial t} + \mathbf{U}.\nabla \ . \tag{23.12}$$

For instance, the material derivative of fluid density ρ is

$$\frac{D\rho}{Dt} \equiv \frac{\partial \rho}{\partial t} + U_i \frac{\partial \rho}{\partial x_i} = \frac{\partial \rho}{\partial t} + \mathbf{U}.\nabla \rho \ . \tag{23.13}$$

23.12 Tensors

Given a coordinate basis or fixed frame of reference, a tensor can be represented as an organized multi-dimensional array of numerical values. The order (also degree or rank) of a tensor is the dimensionality of the array needed to represent it, or equivalently, the number of indices needed to label a component of that array. For example, a linear map is represented by a matrix (a two-dimensional array) in a basis and therefore is a second-order tensor. A vector is represented as a one-dimensional array in a basis and is a first-order tensor. Scalars are single numbers and are thus zeroth-order tensors. Because tensors express a relationship between vectors, tensors themselves must be independent of a particular choice of coordinate system. The coordinate independence of a tensor then takes the form of a transformation law that relates the array computed in one coordinate system to that computed in another one.

A *zeroth-order tensor* is a scalar. It has $3^0 = 1$ component, which has the same value in every coordinate system. Examples are physical quantities such as density, temperature, and pressure.

A *first-order tensor*, e.g. $\mathbf{u}$ [−], has $3^1 = 3$ components. Some examples include position, velocity, acceleration, and vorticity. In the Cartesian system, the first-order tensor can be represented using unit vectors along the coordinate axes

$$\mathbf{u} = \mathbf{e}_i u_i = \mathbf{e}_1 u_1 + \mathbf{e}_2 u_2 + \mathbf{e}_3 u_3 . \tag{23.14}$$

A *second-order tensor*, $\mathbf{b}$ [−], has $3^2 = 9$ components. An example is the stress tensor. In the Cartesian system, the second-order tensor can be represented using a combination of unit vectors along the coordinate axes

$$\mathbf{b} = \mathbf{e}_i \mathbf{e}_j b_{ij} = \mathbf{e}_1 \mathbf{e}_1 b_{11} + \mathbf{e}_1 \mathbf{e}_2 b_{12} + \mathbf{e}_1 \mathbf{e}_3 b_{13}$$
$$+ \mathbf{e}_2 \mathbf{e}_1 b_{21} + \mathbf{e}_2 \mathbf{e}_2 b_{22} + \mathbf{e}_2 \mathbf{e}_3 b_{23}$$
$$+ \mathbf{e}_3 \mathbf{e}_1 b_{31} + \mathbf{e}_3 \mathbf{e}_2 b_{32} + \mathbf{e}_3 \mathbf{e}_3 b_{33} . \tag{23.15}$$

Note that for the second-order tensors, the Einstein's notation requires that the summation over repeated indices be expanded in two dimensions. In the formula above, since both indices i and j have been repeated, the summation requires nine terms over two dimensions to include all combinations of indices [1–3].

Problems

23.1 How many components are there for a *third-order tensor*?

23.2 A velocity vector field is given by the following expression. Calculate the curl of this velocity vector, i.e. $\nabla \times \mathbf{U}$ [s^{-1}]. Also evaluate the curl at the point given by the coordinates $x = 1$, $y = 2$, and $z = 3$ [m].

$$\mathbf{U} = xyz\mathbf{i} + x^2 y^2 z^2 \mathbf{j} + x^{-1} y^{-1} z^{-1} \mathbf{k} . \tag{23.16}$$

23.3 An environmental engineer is analysing a transient velocity field in a fluid flow given by

$$\mathbf{U} = -3xt\mathbf{i} - 3yt\mathbf{j} + 6zt\mathbf{k} , \tag{23.17}$$

where x, y, and z [m] are the cartesian coordinates and t [s] is the time. In addition, the temperature field in this flow is given by

$$T = e^{-t}(x + y + z) \ . \tag{23.18}$$

Help this engineer find an expression for the material derivative of temperature in this velocity field, as a function of x, y, z [m], and t [s], given by

$$\frac{DT}{Dt} = \frac{\partial T}{\partial t} + \mathbf{U}.\nabla T \ . \tag{23.19}$$

References

1. Acheson D J (1990) Elementary fluid dynamics. Oxford University Press, Oxford
2. Pope S B (2000) Turbulent flows. Cambridge University Press, Cambridge
3. Davidson P A (2005) Turbulence: an introduction for scientists and engineers. Oxford University Press, Oxford

Chapter 24
Numerical Methods

Abstract This chapter briefly introduces basic topics in numerical methods that assist the study of turbulence. These methods help discretizing and solving transport equations for fluids using digital computers. The topics covered include Taylor series expansion, the finite difference method, the Newton's method for solving non-linear system of equations, the explicit and implicit Euler methods, and under relaxation.

24.1 Taylor Series Expansion

Any continuous differentiable function $\phi(x)$ [−] can, in the vicinity of x_i [−], be expressed as a *Taylor series*

$$\phi(x) = \phi(x_i) + (x - x_i)\left(\frac{\partial \phi}{\partial x}\right)_i + \frac{(x - x_i)^2}{2!}\left(\frac{\partial^2 \phi}{\partial x^2}\right)_i$$

$$+ \frac{(x - x_i)^3}{3!}\left(\frac{\partial^3 \phi}{\partial x^3}\right)_i + \ldots + \frac{(x - x_i)^n}{n!}\left(\frac{\partial^n \phi}{\partial x^n}\right)_i + H.O.T. , \qquad (24.1)$$

where $H.O.T.$ [−] means *higher order terms*. The Taylor series expansion is at the core of numerical methods [1, 2].

24.2 The Finite Difference Method

In finite difference methods, the derivatives in transport equations are replaced by finite differences. This is supported by the Taylor series expansion, which can be rearranged to give a derivative having differences and derivatives of other order. The core idea in finite difference methods is to approximate a derivative by finite differences and ignore other order derivatives [2]. Suppose a problem is discretized with $\Delta x = x_{i+1} - x_i$ [−] for all i. The first-order derivative can then be approximated using

$$\left(\frac{\partial \phi}{\partial x}\right)_i \approx \frac{\phi_{i+1} - \phi_i}{x_{i+1} - x_i} \;, \tag{24.2}$$

$$\left(\frac{\partial \phi}{\partial x}\right)_i \approx \frac{\phi_i - \phi_{i-1}}{x_i - x_{i-1}} \;, \tag{24.3}$$

$$\left(\frac{\partial \phi}{\partial x}\right)_i \approx \frac{\phi_{i+1} - \phi_{i-1}}{x_{i+1} - x_{i-1}} \;, \tag{24.4}$$

which are known in order as, *Forward-Difference Scheme (FDS)*, *Backward-Difference Scheme (BDS)*, and *Central-Difference Scheme (CDS)*.

The same technique can be used to approximate second derivatives. For instance, for equidistant spacing, the second derivative can be approximated using central differencing such that

$$\left(\frac{\partial^2 \phi}{\partial x^2}\right)_i \approx \frac{\phi_{i+1} - 2\phi_i + \phi_{i-1}}{(\Delta x)^2} \;. \tag{24.5}$$

More complex second-order derivatives can be approximated in situations where inner and outer derivatives exist with some variable inside the outer derivative beside the inner derivative. For example,

$$\left[\frac{\partial}{\partial x}\left(\Gamma \frac{\partial \phi}{\partial x}\right)\right]_i \approx \frac{\left(\Gamma \frac{\partial \phi}{\partial x}\right)_{i+\frac{1}{2}} - \left(\Gamma \frac{\partial \phi}{\partial x}\right)_{i-\frac{1}{2}}}{\frac{1}{2}(x_{i+1} - x_{i-1})}$$

$$\approx \frac{\Gamma_{i+\frac{1}{2}} \frac{\phi_{i+1} - \phi_i}{x_{i+1} - x_i} - \Gamma_{i-\frac{1}{2}} \frac{\phi_i - \phi_{i-1}}{x_i - x_{i-1}}}{\frac{1}{2}(x_{i+1} - x_{i-1})} \;, \tag{24.6}$$

where Γ [−] is a variable and has to be evaluated at half-integer indices, i.e. $i+\frac{1}{2}$ and $i - \frac{1}{2}$. To simplify the approximation, the inner derivatives have been approximated using forward and backward schemes [2].

24.3   Newton's Method for Solving Non-linear System of Equations

The master method in solving non-linear equations or a system of equations is the *Newton's method*. This method is also called the *linearization method*, in which a non-linear function at a point can be replaced by the terms of Taylor series that only contain constants or first-order derivatives.

For a uni-variable non-linear function $f(x)$ [−], the value of the function around x_0 [−] can be replaced by

$$f(x) \approx f(x_0) + \left(\frac{df}{dx}\right)_{x_0} (x - x_0) . \tag{24.7}$$

For instance if $f(x) = x^2$ [−], then $f(x) \approx x_0^2 + 2x_0(x - x_0) = -x_0^2 + 2x_0x$ [−], which is a linear function. This method is extremely useful since it can convert any non-linear system of equations into a linear system of equations. The effectiveness of this method for solving a non-linear system of equations depends on other properties of the non-linear system.

The Newton method can be applied to multi-variable non-linear functions as well. For instance, if f [−] is a non-linear function of three variables x, y, and z [−], then the linearized form of f [−] around x_0, y_0, and z_0 [−] is

$$\begin{aligned}
f(x, y, z) \approx f(x_0, y_0, z_0) &+ \left(\frac{\partial f}{\partial x}\right)_{x_0, y_0, z_0} (x - x_0) \\
&+ \left(\frac{\partial f}{\partial y}\right)_{x_0, y_0, z_0} (y - y_0) \\
&+ \left(\frac{\partial f}{\partial z}\right)_{x_0, y_0, z_0} (z - z_0) ,
\end{aligned} \tag{24.8}$$

where the derivatives have now been replaced by partial derivatives. This method can also be extremely effective in reducing many highly non-linear systems of equations to a linear system of equations [2].

24.4 Explicit and Implicit Euler Methods

Many transport problems are transient in nature. That is a solution of interest is sought for not only in the spatial domain but also in the temporal domain. Therefore, numerical methods must be developed to solve a transport equation in both space and time. The Euler's method is the most common approach for this purpose, and there are two classes of Euler's method: explicit and implicit. We demonstrate these methods using one-dimensional transport equation for a scalar ϕ [−], e.g. concentration of a species in a fluid domain, with constant fluid velocity U [ms^{-1}]. The transport equation is given as

$$\underbrace{\frac{\partial \phi}{\partial t}}_{\text{Storage}} = \underbrace{-U\frac{\partial \phi}{\partial x}}_{\text{Advection}} + \underbrace{\Gamma\frac{\partial^2 \phi}{\partial x^2}}_{\text{Diffusion}} . \tag{24.9}$$

Consider that a *Central-Difference Scheme (CDS)* is used to discretize the equation using the finite difference method. Assume that ϕ_i^n [−] represents the solution in spatial location i [−] and time step n [−] and that the spatial and temporal domains are discretized by Δx [m] and Δt [s], respectively. The finite difference version of the transport equation can be shown using

$$\frac{\phi_i^{n+1} - \phi_i^n}{\Delta t} = -u \frac{\phi_{i+1}^n - \phi_{i-1}^n}{2\Delta x} + \Gamma \frac{\phi_{i+1}^n - 2\phi_i^n + \phi_{i-1}^n}{(\Delta x)^2} \,, \tag{24.10}$$

where the spatial derivatives are discretized at time step n [−]. This equation can be simplified by grouping solutions at a particular spatial location and time step to arrive at the following equation:

$$\phi_i^{n+1} = \left(d + \frac{c}{2}\right) \phi_{i-1}^n + (1 - 2d)\phi_i^n + \left(d - \frac{c}{2}\right) \phi_{i+1}^n \,, \tag{24.11}$$

where we have introduced the dimensionless parameters

$$\begin{cases} d = \frac{\Gamma \Delta t}{(\Delta x)^2} \,, \\ c = \frac{u \Delta t}{\Delta x} \,. \end{cases}$$

The parameter d [−] is the ratio of time step Δt [s] to the characteristic diffusion time $(\Delta x)^2 / \Gamma$ [s], which is roughly the time required for a disturbance to be transmitted by diffusion over distance Δx [m]. The second quantity, parameter c [-], is the ratio of the time step Δt [s] to the characteristic convection time $\Delta x / u$ [s], the time required for the disturbance to be convected a distance of Δx [m]. This ratio is called the *Courant number*

$$Co = \frac{u \Delta t}{\Delta x} \,. \tag{24.12}$$

The equations over all spatial locations can be combined so that the set of equations can be written in the matrix form

$$\boldsymbol{\phi}^{n+1} = \mathbf{A} \boldsymbol{\phi}^n \,, \tag{24.13}$$

where the boundary conditions can be absorbed in the system of equations as well. This suggests that to solve the transient transport equation one needs to choose an initial solution ϕ^0 [−] and then repetitively perform the matrix multiplication over a finite number of time steps to obtain new solutions ϕ^n [−]. This is called the *explicit Euler method*. As convenient as it appears, this method is severely limited because it necessitates the discretization to be extremely refined. So the method is only *conditionally stable* if Δx [m] and Δt [s] meet specific criteria. For instance, for flows that are dominated by convection, where diffusion is negligible, such as atmospheric flows, the Courant number has to be less than one, i.e.

$$Co = \frac{u \Delta t}{\Delta x} < 1 .$$

(24.14)

Alternatively, the transport equation can be discretized in such a way that the spatial derivatives are discretized at time step $n + 1$, i.e.

$$\frac{\phi_i^{n+1} - \phi_i^n}{\Delta t} = -u \frac{\phi_{i+1}^{n+1} - \phi_{i-1}^{n+1}}{2\Delta x} + \Gamma \frac{\phi_{i+1}^{n+1} - 2\phi_i^{n+1} + \phi_{i-1}^{n+1}}{(\Delta x)^2} .$$

(24.15)

This equation can be simplified by grouping solutions at a particular spatial location and time step to arrive at the following equation:

$$\left(-\frac{c}{2} - d\right) \phi_{i-1}^{n+1} + (1 + 2d)\phi_i^{n+1} + \left(\frac{c}{2} - d\right) \phi_{i+1}^{n+1} = \phi_i^n .$$

(24.16)

The resulting system of equations can be written in matrix form such that

$$\mathbf{A}\phi^{n+1} = \phi^n .$$

(24.17)

Now it appears that to arrive at the solution for a new time step, not only one must have a solution in the previous time step, but also one needs to solve a linear system of equations. This is the *implicit Euler method*. It happens that this method results in better *stability* and poses less restrictions on resolution of the discretization, i.e. the magnitude of Δx [m] and Δt [s] [2].

24.5   Under Relaxation

When solving a system of equations iteratively, it is sometimes more stable to only partially update a solution after each iteration. This is known as *under relaxation*. Consider that ϕ^{n-1} [−] is the solution space found in the previous iteration and ϕ^{new} [−] is the newly found solution. With the *under relaxation factor* $0 < \alpha < 1$ [−], the solution can be updated for the next iteration such that

$$\phi^n = \phi^{n-1} + \alpha(\phi^{new} - \phi^{n-1}) .$$

(24.18)

Particularly, whenever solving a non-linear system of equations, this method improves stability of obtaining a numerical solution [2].

Problems

24.1 A continuous differentiable 1D function $\phi(x)$ [−] is being considered that only depends on x [−] as an independent variable. We like to approximate the first- and second-order derivatives of $\phi(x)$ [−], i.e. $\left(\frac{\partial\phi}{\partial x}\right)_i$ and $\left(\frac{\partial^2\phi}{\partial x^2}\right)_i$ [−] at point i [−], using the finite difference method. Consider a uniform discretization of $\Delta x = x_{i+1} - x_i = 2$ [−], given any i [−]. Suppose the values of ϕ [−] at points $i = 1, 2$, and 3 [−] are given as 2, 4, and 7 [−], respectively. The function ϕ [−] is unitless. Show that the first derivative of ϕ [−] at point $i = 2$ [−] using the Forward-Difference Scheme (FDS), Backward-Difference Scheme (BDS), and Central-Difference Scheme (CDS) is approximated, respectively, by

$$\left(\frac{\partial\phi}{\partial x}\right)_{i=2} \approx 1.5, 1, 1.25 . \tag{24.19}$$

Also show that the second-order derivative of ϕ [−] at point $i = 2$ [−] using the Central-Differencing Scheme (CDS) is approximated by

$$\left(\frac{\partial^2\phi}{\partial x^2}\right)_{i=2} \approx 0.25 . \tag{24.20}$$

24.2 A continuous differentiable 1D function $\phi(x) = e^x$ [−] is being considered that only depends on x [−] as an independent variable. This function is to be approximated using the first three terms of its Taylor series expansion around point $x = 1$ [−]. Show that the approximation as a function of x [−] is given by

$$\phi(x) \approx \left(x + \frac{(x-1)^2}{2}\right) e . \tag{24.21}$$

Now use this approximation to evaluate the function at $x = 1.5$ [−]. How close is this approximation to the true value of the function, i.e. $\phi(x = 1.5)$ [−]?

24.3 A mathematician wishes to approximate a continuous differentiable 1D function $\phi(x) = \sin x$ [−], that only depends on x [−] as an independent variable, by the first three terms of its Taylor series expansion around point $x = 1$ [−]. (a) What is the functional form of this approximation, i.e. as a function of x [−]? (b) Now use this approximation to evaluate the function at $x = 1.2$ [−]. How close is this approximation to the true value of the function, i.e. $\phi(x = 1.2)$ [−]?

References

1. Pope S B (2000) Turbulent flows. Cambridge University Press, Cambridge
2. Ferziger J H, Perić M (2002) Computational methods for fluid dynamics. Springer, Berlin

Index